PHalarope Books

PHalarope books are designed specifically for the amateur naturalist. These volumes represent excellence in natural history publishing. Each book in the PHalarope series is based on a nature course or program at the college or adult education level or is sponsored by a museum or nature center. Each PHalarope book reflects the author's teaching ability as well as writing ability.

BOOKS IN THE SERIES:

The Curious Naturalist
John Mitchell and the
Massachusetts Audubon Society

The Amateur Naturalist's Handbook
Vinson Brown

At the Sea's Edge:
An Introduction to Coastal Oceanography
for the Amateur Naturalist
William T. Fox
Illustrated by Clare Walker Leslie

Suburban Wildlife:
An Introduction to the Common Animals
of Your Back Yard and Local Park
Richard Headstrom

Suburban Wildflowers:
An Introduction to the Common Wildflowers
of Your Back Yard and Local Park
Richard Headstrom

The Wildlife Observer's Guidebook
Charles E. Roth
Massachusetts Audubon Society

The Plant Observer's Guidebook:
A Field Botany Manual
for the Amateur Naturalist
Charles E. Roth
Massachusetts Audubon Society

Exploring Tropical Isles and Seas:
An Introduction for
the Traveler and Amateur Naturalist
Frederic Martini

The Indoor Naturalist:
Observing the World of Nature
Inside Your Home
Gale Lawrence

The Seaside Naturalist

A Guide to Nature Study at the Seashore

Written and Illustrated by

Deborah A. Coulombe

in cooperation with the University of New Hampshire

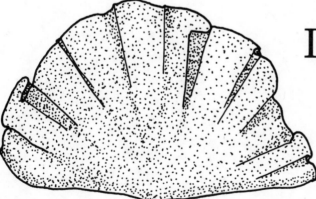

PRENTICE HALL PRESS • NEW YORK

To my parents

*The Seaside Naturalist: A Guide to Nature Study at
the Seashore* is adapted from the *Tidepool Times*, newsletter
of the Nature Center, Odiorne Point State Park.
The Nature Center at Odiorne Point State Park, Rye,
New Hampshire, is sponsored by the New Hampshire
Division of Parks and Recreation,
the Audubon Society of New Hampshire, and the
University of New Hampshire/University of Maine
Sea Grant College Program.

Published in 1987 by Prentice Hall Press
A Division of Simon & Schuster, Inc.
Gulf + Western Building
One Gulf + Western Plaza
New York, NY 10023

Originally published by Prentice-Hall, Inc.

PRENTICE HALL PRESS is a trademark of Simon & Schuster, Inc.

Library of Congress Cataloging-in-Publication Data

Coulombe, Deborah A.
   The seaside naturalist.

   (PHalarope books)

   Bibliography: p.
   Includes index.
   Summary: An illustrated guide to the characteristics of some of the 200,000
different plants and animals that live in the ocean.

   1. Seashore biology. [1. Seashore biology. 2. Marine animals. 3. Marine plants]
I. Title.
QH95.7.C68 1984   574.909′46   84-1972
ISBN 0-13-797259-8
ISBN 0-13-797242-3 (pbk.)

Manufactured in the United States of America

10   9   8   7   6   5   4

First Prentice Hall Press Edition

Contents

Foreword

After most of the winter winds die down, the spring migration begins. If you come to Odiorne Point State Park in May or June when the tide is low you will see them: scattered groups of adults and children working knee deep in the tidal pools of this seaside park. From a distance they resemble multi-colored gulls, their shouts and chatter and laughter mingling with the splash of the waves on the shore. They are an ungraceful lot, slipping and sliding as they work among the rocks, looking, questioning, explaining, exploring, touching gently and replacing everything where it was found. In the summer the posture is the same, though the plumage changes. Sunglasses and zinc oxide noses gradually replace the mittens and woolens and winter jackets worn during the chill New Hampshire spring. No matter what the season, there is a strong facination with the rocky shores and salt marshes of Odiorne Point.

It was because of that kind of interest that the Nature Center at Odiorne was established under the sponsorship of the Audubon Society of New Hampshire, the New Hampshire Division of Parks, and the University of New Hampshire Marine and Sea Grant programs. Gradually exhibits and programs were developed to interpret the resources of the park. Though small, the Nature Center quickly became an important resource for the community.

Several years ago, when the marine awareness programs first began, a teacher from nearby Hampton, New Hampshire, roared into the Center waving a newsletter. She had just returned from visiting several other environmental education centers and came back to New Hampshire loaded with information to share. Her most important message was, "We need a newsletter for the Nature Center at Odiorne," a splendid idea, but where could a talented writer and artist be found to produce such a publication?

Coincidentally, within two weeks Deborah Coulombe contacted me. She had worked at the Nature Center as a student intern the year before and would be there again as a summer naturalist. She sent some samples, saying, "I've got this great idea for a newsletter. . . ." The *Tidepool Times* became a perfect teaching tool, an adjunct to the Center's lectures, workshops, and programs.

It provided that extra spark to help our visitors to focus in on the environment . . . up close. To help them look for details, to see similarities and differences. To stimulate their curiosity, getting them to wonder about where something lived or why it was built a certain way. Most importantly it has helped us to get people to think about relationships: how we affect the environment and how it affects us. The bottom line, however, is decision making. The *Tidepool Times* is helping us to make people more aware of their environment so that they will become concerned and ultimately they will be motivated to act, motivated to make better decisions about its future.

In New Hampshire we have just eighteen miles of open coast, so our marine resources are very precious and in great demand. For us careful decision making is becoming more and more critical. However, these challenges to the marine environment do not stop at New Hampshire's borders. All of the coastal communities in the east are confronted with issues dealing with the utilization of their various marine resources, which is why it has been an exciting time to watch our newsletter, the *Tidepool Times,* blossom into *The Seaside Naturalist.*

No matter where the location, there are no easy solutions to resolving coastal issues and the conflicting problems of multiple use, nor will *The Seaside Naturalist* attempt to provide those answers. However, *The Seaside Naturalist*

can help to heighten our awareness so that ultimately we *are prepared* to make better choices. Choices that will make sure that there is something left for future generations to make wise decisions about.

Julia Steed Mawson
Nature Center Director

Preface

I hope *The Seaside Naturalist* is a valuable resource for those people who want a book combining some of the features of a field guide with some of the concepts of a marine biology text. *The Seaside Naturalist* in no way attempts to identify all the organisms on the Atlantic Coast or discuss biological concepts in depth. Instead, I have tried to answer some of the questions that educators, beachgoers, and incipient marine biologists have asked of me in my experience as a marine naturalist.

Originally, *The Seaside Naturalist* was intended to be a compilation of articles from the *Tidepool Times*, the marine education newsletter of the Nature Center at Odiorne Point State Park, in Rye, New Hampshire. With funding from the University of New Hampshire/University of Maine Sea Grant Program, the New Hampshire Division of Parks, and the Audubon Society of New Hampshire and the guidance of Julia Steed Mawson of the Office of Marine Extension and Public Education at UNH, I have been writing and illustrating the *Tidepool Times* for several years. Many ideas from the *Tidepool Times* do appear in *The Seaside Naturalist;* however, most of the text and illustrations are new.

For sustaining and encouraging my interest in science and education, I give thanks to three fine organizations for whom I have had the pleasure of working:

The Chesapeake Bay Center for
  Environmental Studies
The Smithsonian Institution
Edgewater, Maryland

The University of New Hampshire
  Marine Program
University of New Hampshire
Durham, New Hampshire

Newfound Harbor Marine Institute
  Big Pine Key, Florida

Special thanks to my fellow marine educators who all contributed to this book in their own way: Donna Stewart for her artistic suggestions; Pam Miller for her psychological expertise and philosophical insights; Erick Lindblad for his patience and flexibility; Jim and Sharon Thomas for their psychological assistance; Chelsie Wallace, Don Levitan, Brad Baldwin, Ginger Eisenman, Lisa Gabriel, and Denny Lane for reviewing my first draft. For entertaining me through arduous hours of typing, I thank Jeff Holmquist and Rick Maddox.

A special thanks to Mary Kennan at Prentice-Hall for originating the idea of this book and having the patience to work with a novice author.

A Note on Scientific Names

Each living organism is assigned a scientific name consisting of two parts: the generic name and the specific name. This is known as the *system of binomial nomenclature* and was first developed by Linnaeus in 1758. The generic name is capitalized, the specific name is not; both are either italicized or underlined. Each type of organism has only one specific name, and it is (usually) accepted worldwide. In this book, common names and scientific names are given. Although the temptation is strong to "bleep" over the scientific name as one would over names in a Russian novel, it is worth the time to try to pronounce and/or translate them. Jaeger's *Sourcebook of Biological Names and Terms,* published by Charles C Thomas, can help with that.

Species are grouped together in genera (singular: genus). Genera are grouped together in families; families are grouped in orders, orders in classes, classes in phyla, and phyla in kingdoms. In the larger phyla, there are many intermediate categories. Look at the classification of the Rock crab, *Cancer irroratus:*

Kingdom Animalia
Phylum Arthropoda
Subphylum Crustacea
Class Malacostraca
Order Decapoda
Suborder Pleoyemata
Infraorder Brachyura
Section Cancridea
Family Cancridae
Genus *Cancer*
Species *Irroratus*

Acknowledgment is made to the following for the granting of permission to reprint adaptations of drawings from their publications:

The drawing on page 6 is adapted from *Introductory Oceanography* (2nd ed.) by H. V. Thurman; published by Charles E. Merrill Publishing Co. Copyright © 1978.

Drawings on pages 57, 71, 116, and 118 are adapted from *Invertebrate Zoology* (4th ed.) by Robert D. Barnes. Copyright © 1980 by Saunders College/Holt, Rinehart and Winston, CBS College Publishing.

Drawings on pages 78, 79, and 87 are adapted from *A Field Guide to the Atlantic Seashore* by Kenneth L. Gosner. Copyright © 1978 by Kenneth L. Gosner. Used by permission of Houghton Mifflin Company.

The illustrations on page 113 are adapted from *Seashells of North America,* which is illustrated by George F. Sandstrom.

The drawings on page 122 are adapted from *Mollusks* by Paul Bartsch, Dover Publications Inc., New York, 1968.

The drawings on pages 148 and 157 are adapted (with permission) from *Seashore Life of Florida and the Caribbean* by Gilbert Voss. Published by Banyan Books, Miami.

The drawings on pages 164, 166, 169, 170, 177, 178, 179, and 181 are adapted from *Fishes of the Gulf of Maine* by Bigelow and Schroeder, Harvard University, 1953.

The drawing of the whale shark that appears on page 182 is redrawn from Castro and Stone's *Sharks of North America,* Texas A & M University Press, 1983. Used by permission.

Page 187's illustrations are adapted from art done by Mary Beath for the Center for Environmental Education, Washington D.C.

The drawings on page 207 are adapted from Watson and Ritchie's *Sea Guide to Whales of the World* (E.P. Dutton, New York). Published by Bellen and Higton, a part of the Hutchinson Publishing Group.

The drawings on page 209 are adapted with permission from drawings by D.D. Tyler and John R. Quinn that appear in *A Field Guide to the Whales, Porpoises, and Seals of the Gulf of Maine and Eastern Canada* (3rd ed.) by Katona and Richardson. Published by Charles Scribner's Sons.

Introduction to the Ocean

If the earth were the size of a basketball, the ocean would be a mere film of water over its surface. Actually, the ocean is a layer of water with an average depth of 2.3 miles on the surface of a sphere 8000 miles wide.

The surface of the earth is nearly 200 million square miles. Over this surface, continents and oceans are unevenly distributed. Two-thirds of the earth's land area lies in the Northern Hemisphere, while the Southern Hemisphere contains eighty percent of the water.

Today, seventy-one percent of the earth's surface is covered by the ocean. During past geologic eras however, the oceans may have covered slightly less or more of this surface depending on sea level. A decrease in sea level (perhaps due to an Ice Age) would have resulted in more of a continent being above water, while a rise in sea level would have the opposite effect. Tradionally, the ocean has been separated into separate basins: the Pacific, Atlantic, Indian and Arctic. The ocean is actually one large interconnecting system constantly mixed by winds, tides currents, waves and upwelling.

Most people on earth live within a few hundred miles of the ocean. Twenty-three of the fifty United States are on the seacoast. Excluding Alaska and Hawaii, the United States has nearly 5000 miles of shoreline. New Hampshire accounts for 18 of these miles; Florida for 1350. The shoreline of the eastern United States ranges from the rocky boreal coast of Maine to the sub-tropical, Caribbean-influenced shore of the Florida Keys. Between these two extremes lie a variety of different marine habitats: salt marshes, sandy beaches, eel grass beds, and mangrove islands, to name a few. The ocean is the common link between these diverse shorelines and communities.

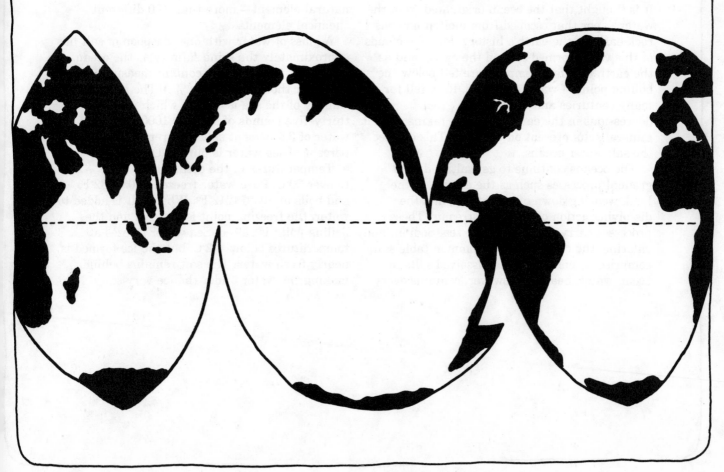

All about Sea Water

dissolved salts in sea water

It is thought that the ocean originated from the water vapor that escaped from molten igneous rock early in the earth's history. Massive clouds of this vapor formed around the earth, and as the earth's surface temperature fell below the boiling point of water, it rained. Rain fell for many centuries and drained into great depressions in the earth's surface, taking with it mineral salts present on the land. This created the salt water oceans.

The oceans continue to get saltier due to gradual processes such as the erosion of the land, wearing down of mountains, and the dissolving action of rain and streams. These processes carry minerals to the sea. Sodium and chlorine, the components of common table salt, comprise 85 percent of the dissolved salts in ocean water. Sea salt, however, contains every natural element—more than 100 different chemical elements.

A glass of water with one teaspoon of salt has approximately the same salinity as the ocean. Salinity is the total salt content measured in parts per thousand (ppt or ‰). The average salinity of the sea is 35 ppt, which means thirty-five pounds of salt per 1000 pounds of water of 3.5 percent salt. The remaining 96.5 percent of sea water is pure H_2O.

Temperatures in the ocean range from $-2°C$ to over 30°C. Pure water freezes at 0°C (32°F) and boils at 100°C (212°F). When salt is added to water, the freezing point is lowered and the boiling point is raised. Sea water freezes at temperatures below $-2°C$, but the ice formed is nearly fresh water. The salt remains behind, making the water under the ice very salty.

Tides

It was not until Sir Isaac Newton (1642–1727) discovered the law of gravity that the effect of the sun and the moon on the tides was fully understood. Tide-generating forces are a result of the gravitational attraction between the earth, the sun, and the moon. All surfaces of the earth are pulled toward the moon and sun. The land surface, however, is not flexible, while the surface of the ocean is.

As the moon rotates around the earth, tidal bulges are created. Water underneath the moon is pulled toward it, forming a bulge. This bulge of water rotates around the earth beneath the moon (although not *directly* beneath it). On the opposite side of the earth, the centrifugal force resulting from the earth's rotation causes a second bulge. High tides are produced as the bulges hit land and water piles up. Between bulges, low tides occur. Tides occur

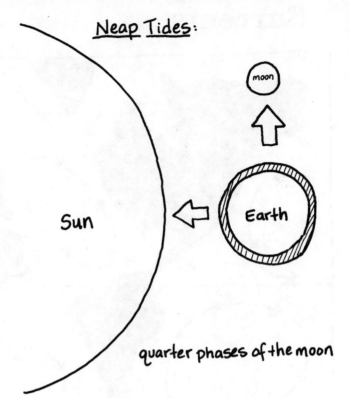

Neap Tides:

quarter phases of the moon

Spring Tides:

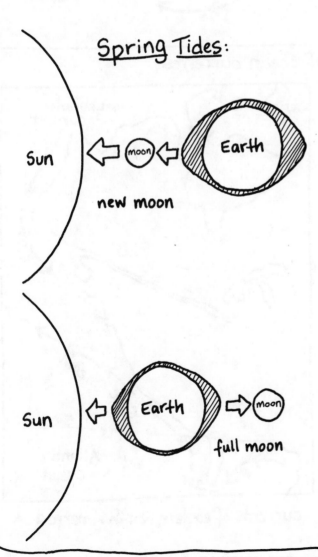

new moon

full moon

approximately fifty minutes later each day because it takes the moon twenty-four hours and fifty minutes to rotate around the earth.

Although the sun is much larger than the moon, it is also much farther away, so that its tide-generating force is only 46 percent that of the moon. During times of the new moon and full moon, the tide-generating forces of the moon and sun are combined. The tidal bulges created by moon and sun align, creating one big bulge. At this time, the vertical difference between high tide and low tide is greatest. This is known as a spring tide.

During the moon's quarter phases, the forces of the sun work at right angles to those of the moon, causing the bulges to cancel each other. This cancelling effect produces a smaller difference between high and low tide and is known as a *neap* tide.

Currents

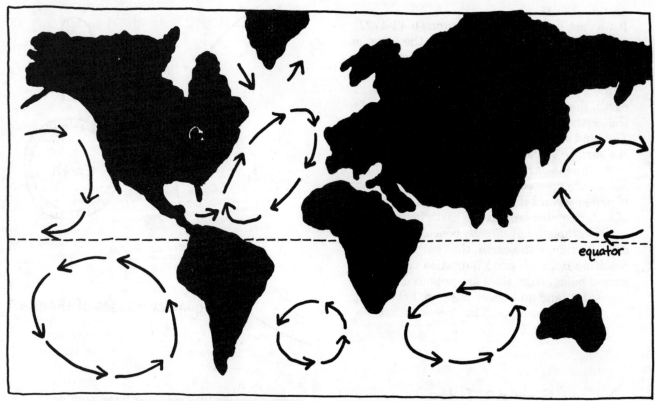

generalized pattern of ocean currents

Ocean waters are always in motion. Surface currents are due primarily to the wind, but tides, rain, evaporation, river runoff, and bottom topography can all affect the movement of the upper ocean water. Winds that drive the surface currents are the Westerlies and the Trade Winds. The Westerlies (40–50° latitudes) blow west to east. The Trade Winds (20° latitudes) blow east to west. These winds are a result of heated air from the tropics traveling to the poles and incorporating the motion of the earth's rotation into their movement. In the Northern Hemisphere, wind-driven currents move clockwise. In the Southern Hemisphere, they move counterclockwise.

Each current has its own characteristic salinity, density, and temperature. The Gulf Stream, which was first mapped by Benjamin Franklin, runs along the east coast of the United States and is one of the strongest currents known. It is a warm, salty current, and although it begins to move offshore north of Cape Hatteras, its warming influence is felt as far north as Cape Cod.

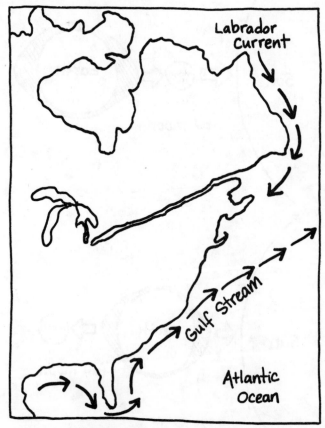

currents of eastern North America

Waves

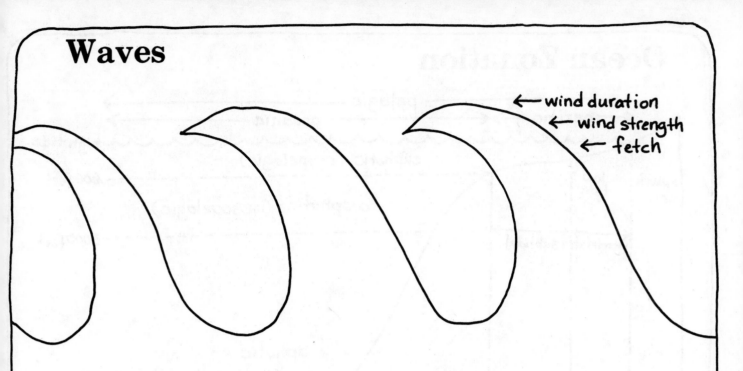

Waves commonly seen on the shore are wind waves. They are caused by the friction of wind blowing across the water. Ripples caused by gentle breezes may play an important role in the formation of these waves. It is thought that the ripples provide the surface roughness necessary for the wind to push or pull the water.

The size of a wave depends on three factors: the distance along open water over which the wind blows (known as the "fetch"), the strength of the wind, and the length of time the wind blows. If all three of these factors are large, the waves will be large.

Although waves are constantly moving and changing, they have definite, measurable features. The *crest* is the highest part of the wave; the *trough* is the lowest. The distance between the crest and the trough is the *wave height*. The distance from crest to crest is the *wave length*. The *period of a wave* is the time it takes for succeeding crests to pass a point. Up to a certain point, waves continue to grow as they absorb energy from the wind. However, when the wave height becomes one seventh the size of the wave length, the wave will topple over in white caps.

The largest wind waves, formed by strong winds blowing over large bodies of water, occur in the open ocean. But as waves advance toward shore, ledges, shoals, coastal islands, and coral reefs cause the higher waves to break. For instance, long swells seldom reach northern New England at full strength because energy is spent as the swells pass over George's Bank.

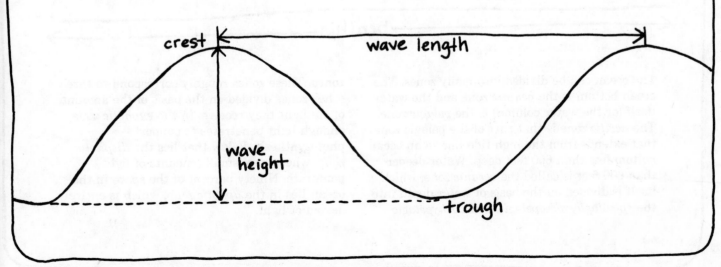

Ocean Zonation

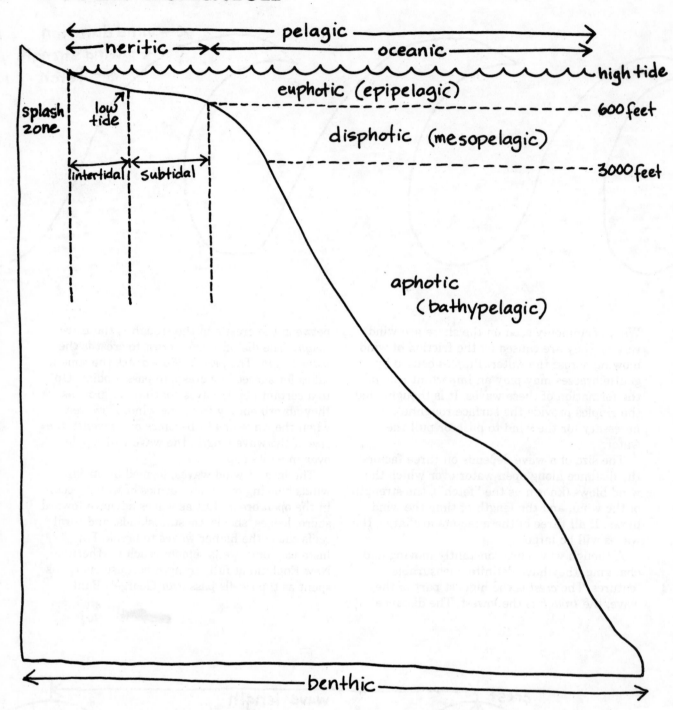

The ocean can be divided into many zones. The ocean bottom is the *benthic* zone and the water itself (or the water column) is the *pelagic* zone. The *neritic* zone is that part of the pelagic zone that extends from the high tide line to an ocean bottom less than 600 feet deep. Water deeper than 600 feet is called the *oceanic* zone, which itself is divided on the basis of water depth into the *epipelagic, mesopelagic,* and *bathypelagic*

zones. These zones roughly correspond to three other zones divided on the basis of the amount of sunlight they receive. In the *euphotic* zone, enough light penetrates to support photosynthesis. Below that lies the *disphotic* zone, where very small amounts of light penetrate. Ninety percent of the space in the ocean lies in the *aphotic* zone, which is entirely devoid of light.

The Ocean as a Habitat

Over 200,000 different types of plants and animals live in the ocean. These marine organisms have the same four basic needs as terrestrial organisms: water, air, food and a place to live. Compared with the relatively unstable, changeable conditions on land, the sea satisfies these four needs more uniformly because it is a much more stable environment. Marine organisms live surrounded by water, which provides food, air, and support for their bodies.

Water is very important to all types of living organisms; marine plants and animals have an unlimited supply of it. But is is not plain water alone that is essential to metabolic processes; the gases and minerals in sea water are also vital to life.

The Ocean as a Habitat

Atmospheric gases, such as oxygen, nitrogen, and carbon dioxide dissolve into water at the sea surface. The percentages of gases in the ocean differ from the percentages present in the atmosphere, due partly to the solubility of gases in water. For instance carbon dioxide is very soluble in sea water. The concentration of carbon dioxide in sea water is 750 times the concentration of carbon dioxide in the atmosphere.

Gas concentrations differ within the water column itself. Differences in concentrations of gases at certain depths in the sea are partially due to living organisms as well as solubility. Oxygen is produced by plants during photosynthesis, but plants inhabit only the euphotic zone of the ocean. In photosynthesis, plants absorb energy from sunlight and use it to help convert carbon dioxide and water into food. Oxygen is released in the process. Animals use the oxygen for respiration, the process by which they derive their energy. Carbon dioxide is released as a by-product of respiration.

Salts are important in the chemical workings of living cells and are needed by all organisms. Marine plants and animals take salt directly from the water. Salts used in photosynthesis, especially the phosphates and nitrates, are quickly depleted from the euphotic zone, but vertical circulation of the water causes an upwelling of these nutrients from the bottom to the surface. Marine animals use mineral salts in many ways. Calcium is especially important, as mollusks, corals and crustaceans extract it from the water for their shells and skeletons.

Compared with food on land, the food supply in the ocean is relatively dependable and easily obtained. Plants can grow all year round, since water temperature changes are usually not extreme. Animals that do not swim or crawl to obtain food can filter it out of the water as it flows by.

In addition to carrying air, minerals and food, the water supports the bodies of things living in it. Organisms living in sea water tend to be weightless. No energy is wasted fighting gravity. The saltier the water, the more buoyancy it provides.

In the sea, life is found to several miles down. But within this vast area, populations of marine organisms are unevenly distributed. Ninety percent of the sea is in total darkness, yet nearly all marine life depends directly or indirectly on microscopic algae found only in the euphotic zone. Therefore, most of the animals in the ocean live in the euphotic zone, or migrate to it in search of food.

Availability of shelter also governs the distribution of marine life. Areas with many nooks and crevices, such as rocky shores and coral reefs, attract organisms that need places to attach or hide.

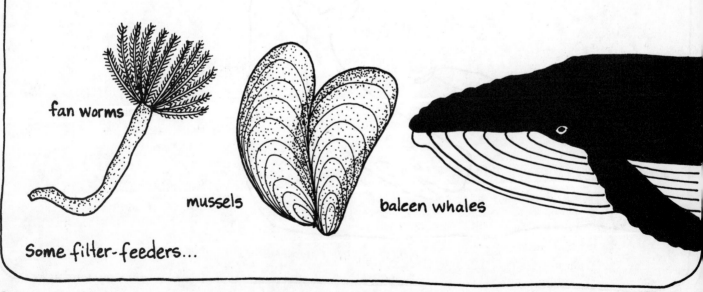

fan worms

mussels

baleen whales

Some filter-feeders...

Ocean True/False Quiz

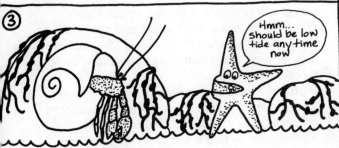

1. Evaporating a cubic foot of seawater will yield 2.2 pounds of salt.

2. At a depth of 10,000 feet, the pressure in the ocean is 500 pounds per square inch.

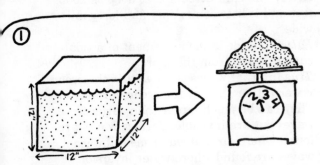

3. Most places along the Atlantic coast have one low and one high tide each day.

4. Tidal waves are caused by tides.

5. The Gulf Stream moves about the same amount of water as the Mississippi River.

6. An ebb tide is the incoming flow of tidal water.

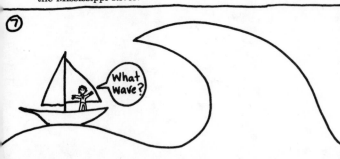

7. The largest ocean waves are invisible.

8. Cold water can hold more dissolved oxygen than warm water.

9. "Boy, I'll tell ya, it's an 'eat and be eaten' world out there."

10. Most sea creatures could live just as easily in fresh water as in salt water.

1. *True* A cubic food of seawater contains 2.2 pounds of salt. All water, even rainwater, contains dissolved salts, although not all water tastes salty. A cubic foot of water from Lake Michigan contains 0.01 pounds of salt (16/100 ounce).

2. *False* The pressure at 10,000 feet is 2.2 *tons* per square inch! The pressure is almost 500 pounds per square inch at 1000 feet.

3. *False* On the Atlantic coast of North America, there are two low tides and two high tides within a period of twenty-four hours and fifty minutes. This is known as a *semi-diurnal* rhythm. Areas in the Gulf of Mexico have one high tide and one low tide a day (a *diurnal* rhythm). The Pacific coast experiences *mixed* tides. There are two low and two high tides a day, but the heights of the two low tides are very different, as are the heights of the two high tides.

4. *False* Tidal waves are caused by sudden movements of the ocean bottom, such as earthquakes, not by tides. Tidal waves are more properly called tsunamis or seismic waves.

5. *False* The Gulf Stream moves about 70,000,000 tons of water per second. That is 1000 times the discharge of the Mississippi River. The Gulf Stream is as deep as 2000 feet and is between twenty and forty miles wide.

6. *False* Ebb tide is the movement of a tidal current *away* from the shore.

7. *True* Internal waves occur *within* the ocean water rather than at the ocean surface. They can be much larger than surface waves but are invisible. Surface waves are found where air and water meet. Internal waves are found where water layers of different densities meet.

8. *True* As temperature or salinity increases, oxygen becomes less soluble in sea water.

9. *True* All animals need to eat and many are subsequently eaten themselves. The natural fate of most sea animals is to be eaten by another creature.

10. *False* Most sea creatures are adapted to life in very salty water. They work constantly to maintain the proper salt balance inside themselves. When a sea creature is moved into fresh water, its body fluids are suddenly saltier then the surrounding water. Water rushes into the body to compensate. The cells swell with water and burst. The animal will eventually die.

2 Life in the Sea

Life probably originated in the marine environment. The ocean has supported the evolution of a diversity of organisms, some familiar, others bizarre. All these organisms have adapted to life in the ocean in different ways. Some living things in the ocean look so unlike living things on land that even the basic question, "Is that a plant or an animal?" becomes tricky to answer.

Most plants in the ocean do not have roots, stems or leaves. Many are too small to be seen except under high magnification. Very few bear any resemblance to terrestrial plants. Marine animals are just as unique, and are proof that not all creatures have legs, eyes and ears. Many sea creatures live permanently attached to the bottom and have no need to see, hear or walk. They merely have to remove oxygen and food from the water as it passes by them.

There are definite differences between plants and animals. Understanding these differences is essential to understanding life in the ocean.

All living organisms need food. The basic difference between plants and animals is that plants make their own food, while animals get food from their environment. Through photosynthesis, plants manufacture organic materials (food) from inorganic materials (water, carbon dioxide, and nutrients) using sunlight as their source of energy. Because plants make their own food, they are called *producers*.

Animals are known as *consumers*. They gather and consume organic material rather than making it themselves. *Herbivores* are animals that eat plants. *Carnivores* are animals that eat meat. *Omnivores* are animals that eat both plants and animals. Scavengers eat leftovers and dead organisms. No matter what animals eat, all their food can be traced back to the ability of plants to produce organic material from inorganic substances.

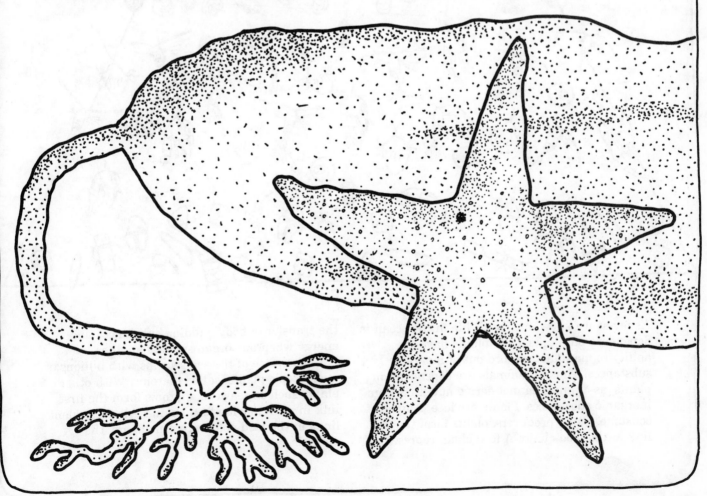

Who Eats What?

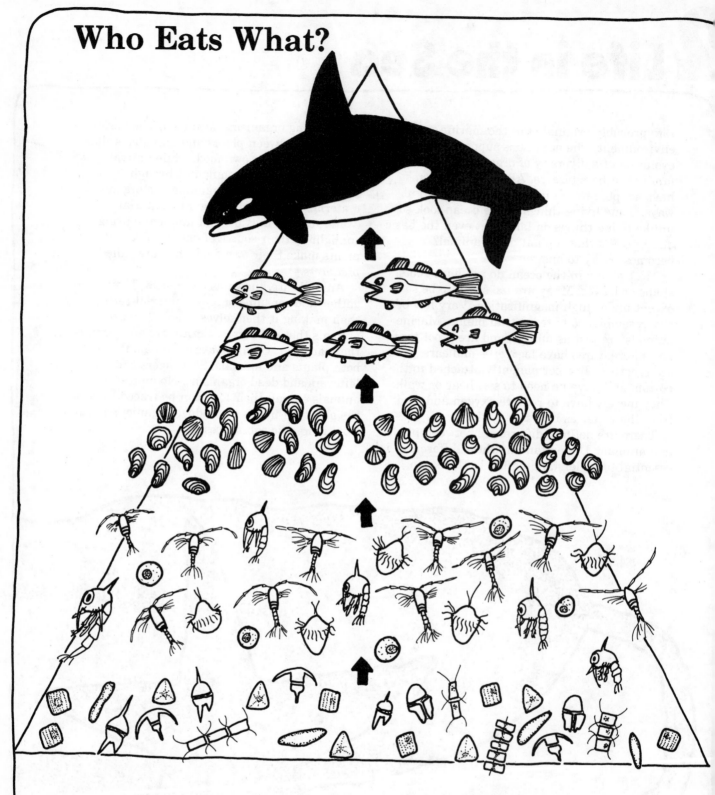

As far as we know, nearly all life in the ocean is dependent on plants. Only plants have the ability to manufacture food out of inorganic substances. Thus all animals are dependent on plants, as animals cannot derive nutrition from inorganic substances. Plants produce. Animals consume. Being producers, plants form the first link in the food chain. A food chain represents the transfer of body-building substances and energy when one organism eats another.

The surface of the sea swarms with billions of microscopic plants, called diatoms. With other plants such as seaweed, diatoms form the first link in most marine food chains. All subsequent links in the food chain are consumers, the animals.

Who Eats What?

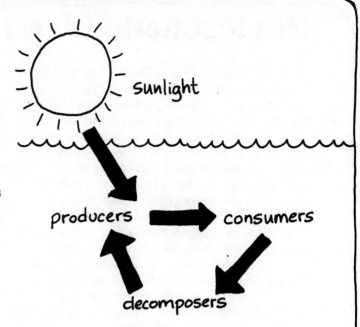

Page 12 shows a marine food chain, with an Orca (Killer whale) as the top carnivore. Orcas often eat cod. Although cod are not finicky eaters, they do have a preference for large bivalves (especially mussels and cockles), which they swallow whole, shell and all. In turn, the bivalves feed on tiny animals that they filter out of the water. These tiny animals (zooplankton) eat even tinier plants: the diatoms and other phytoplankton. Thus the diatoms are the producers. The zooplankton, bivalves, cod, and Orca are all consumers.

In the ocean there are innumerable individual food chains overlapping and intersecting to form complex food webs. Most marine creatures eat a variety of foods. If one link in a chain is depleted, the other consumers in the chain have alternate food sources.

Food chains usually end with an animal nothing else will prey upon, an Orca for example. But eventually that creature dies, and its body slowly sinks to the bottom of the sea. On the way down, and once on the bottom, the dead animal is eaten by scavengers such as sharks, crabs, brittle stars, and lobsters. But the process doesn't end there. Bacteria attack and decompose the remains. As decomposers, bacteria break complex organic substances into simple, inorganic nutrients. Upwelling water currents carry these nutrients to the surface, where they are used by diatoms and other marine plants.

The food potential of an area is directly proportional to the rate at which nutrients are recycled back to the photic zone. Nutrients are renewed quickly where deep waters cycle to the surface, such as shallow regions and areas of upwelling.

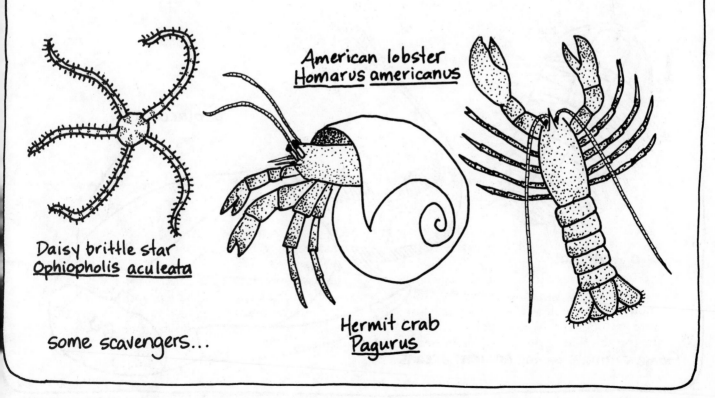

Daisy brittle star
Ophiopholis aculeata

American lobster
Homarus americanus

Hermit crab
Pagurus

some scavengers...

Phylogenetic Classification

As many as five million different kinds of organisms are found on Earth. People have been classifying these organisms for centuries, trying to find order and meaning in their diversity. The classification of living things as producers or consumers is but one way to categorize marine life. Another way is phylogenetically. Phylogenetic classification reflects evolutionary relationships between organisms rather than food relationships.

Life on earth began about three billion years ago. From relatively few primitive forms, the major groups of plants and animals developed. In phylogenetic classification, those organisms with a common evolutionary descent and similar fundamental characteristics are grouped together. The largest and most general phylogenetic subdivision is the *kingdom*. The Plant Kingdom and the Animal Kingdom are the most familiar to us. There are as many as three additional kingdoms, each consisting of organisms that are not clearly plants or animals.

Members of a kingdom are further divided into phyla (singular: phylum), or divisions. Within the Animal Kingdom there are approximately twenty-nine phyla. The apparent origin and relationships of the major phyla of marine animals are shown on the facing page. A small single-celled creature, the protozoan, was probably one of the first animals. Other phyla of animals slowly evolved from that. Organisms that did not successfully adapt to changing environmental conditions have become extinct.

Each phylum is divided into classes. These in turn are followed by other lower divisions: order, family, genus, and species. Each organism has a scientific name consisting of two parts: the generic name, which is always capitalized, and the specific name, which is never capitalized. The two names together comprise the name of the species, thus *Callinectes sapidus,* the Blue crab.

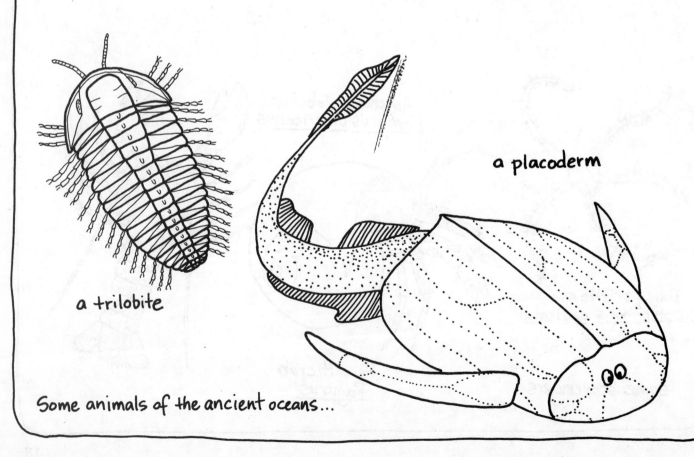

a placoderm

a trilobite

Some animals of the ancient oceans...

Major Groups of Marine Animals and Their Possible Origins

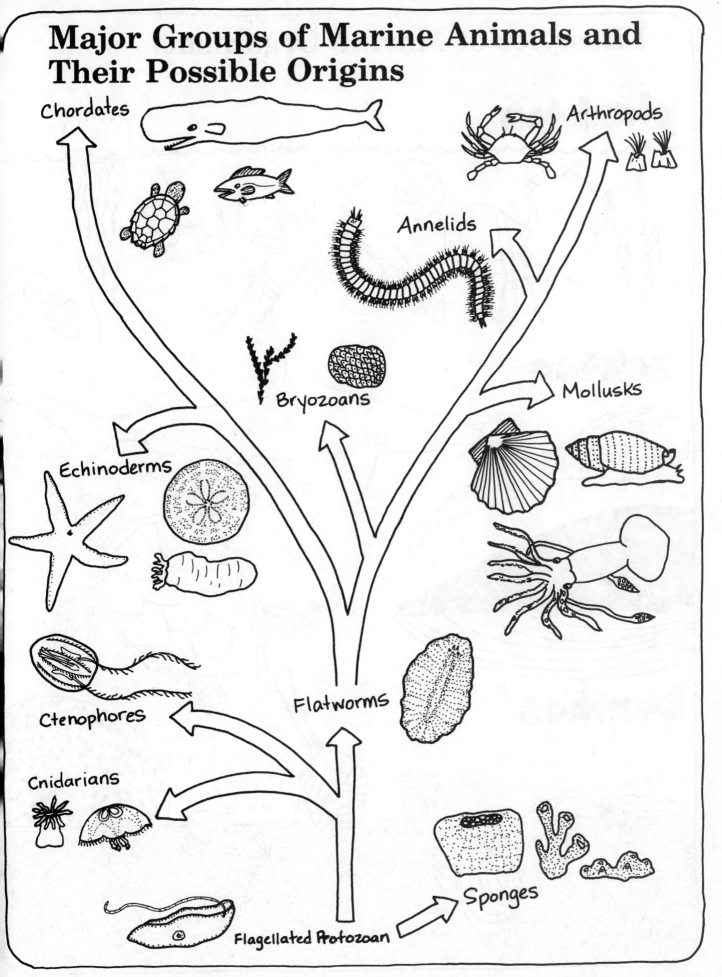

Chordates

Arthropods

Annelids

Bryozoans

Mollusks

Echinoderms

Ctenophores

Flatworms

Cnidarians

Sponges

Flagellated Protozoan

Lifestyles of Marine Organisms

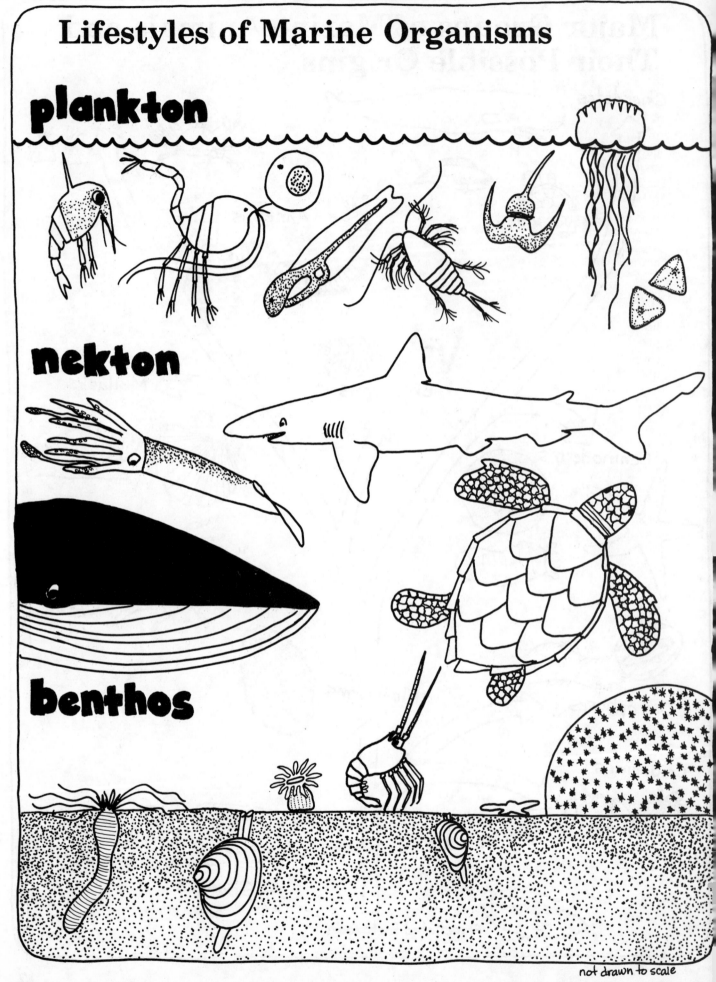

plankton

nekton

benthos

not drawn to scale

Plankton

Marine life can be divided into three categories based on lifestyles. Organisms that live in or on the bottom, such as seaweed or crabs, are called the *benthos*. Strong-swimming animals that live in the open water, such as squid, whales, and adult fish, are called the *nekton*. *Plankton* are small floating or feebly-swimming plants and animals in the water. Plankton may be primitive unicellular organisms or complex multicellular plants and animals. All types of plankton are at the mercy of the waves, tides, and currents for transportation. Most of the organic matter in the sea is plankton, and directly or indirectly, nearly all other marine creatures depend on it as a source of food.

Plant plankton (*phytoplankton*) need to be near the surface, where light is available for photosynthesis. Most animal plankton (*zooplankton*) need to be near the surface to feed upon the phytoplankton. In order to stay afloat near the surface, plankton have evolved many ways to control their position in the sea. Spikes and other projections on a plankter help to distribute the organism's weight over a large surface area, slowing its sinking. Examples are zoea larva and brachiolaria larva. Oil is lighter than water. Many organisms, such as copepods and diatoms, produce oil to help them float. Air-filled floats help many types of marine zooplankton, such as the Portuguese man-o-war, stay afloat.

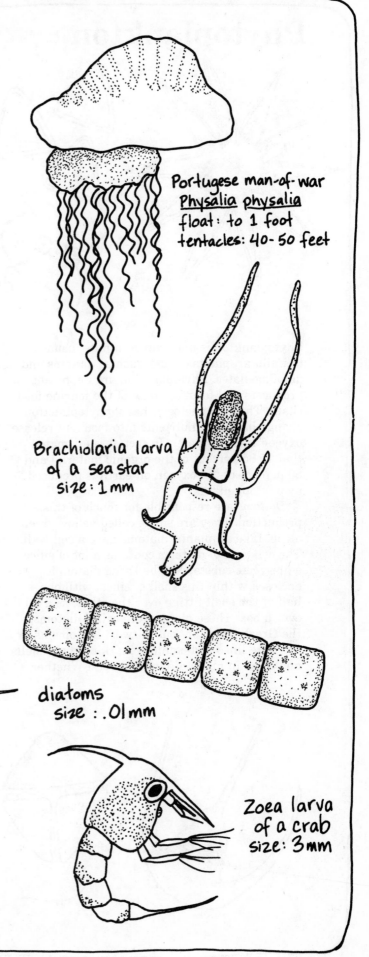

Portugese man-of-war
Physalia physalia
float: to 1 foot
tentacles: 40-50 feet

Brachiolaria larva
of a sea star
size: 1mm

diatoms
size: .01mm

copepods
size: 1.5mm

Zoea larva
of a crab
size: 3mm

Phytoplankton

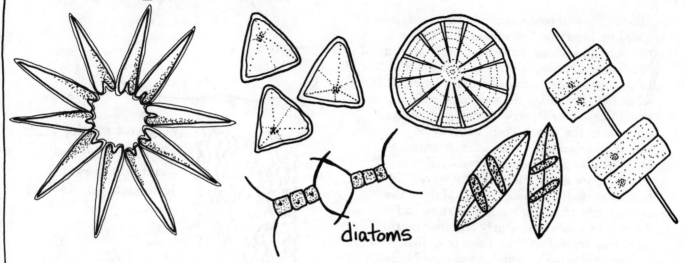

diatoms

Phytoplankton is made up of minute, usually unicellular, plants (algae), such as diatoms and dinoflagellates. Phytoplankton are important producers and form the base of the marine food chain. Through photosynthesis, phytoplankton change inorganic nutrients into food and release oxygen into the water. In fact, it is estimated that 75-85 percent of the organic material and 80 percent of the oxygen on earth is produced by phytoplankton.

Diatoms are responsible for much of this production. They are single-celled yellow-green algae. Like all plants, diatoms have a cell wall. The cell wall of diatoms contains a lot of silica, a glass-like substance. The living diatom is enclosed within this shell of silica, with one–half of the shell fitting over the other, as a lid over a box. These plants were called diatoms, the Greek word for "cut in two." Diatoms have intricate lines and etchings marking their shells and come in many different shapes. Whether alone or in colonies, diatoms are well-adapted for floating, partly due to their shapes and shells. They are probably the single most important food source in the ocean, being eaten not only by small zooplankters, but by larger creatures, such as oysters and clams.

Dinoflagellates resemble both plant and animal plankton. Propelling themselves using two flagella in grooves along their body, dinoflagellates can swim like simple animals. But they can also photosynthesize like plants. This group ranks second to the diatoms in importance as food producers. Two species of dinoflagellates, *Gonyaulax* and *Gymnodinium*, are responsible for the so-called "red tide," the red-colored water that may accompany a sudden appearance of a dense phytoplankton population. These dinoflagellates are responsible for a variety of toxic effects, including fish mortality and paralytic shellfish poisoning.

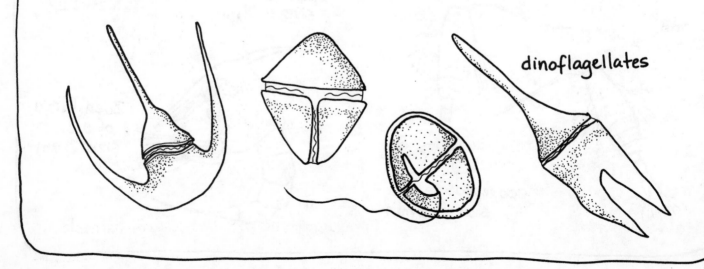

dinoflagellates

Zooplankton

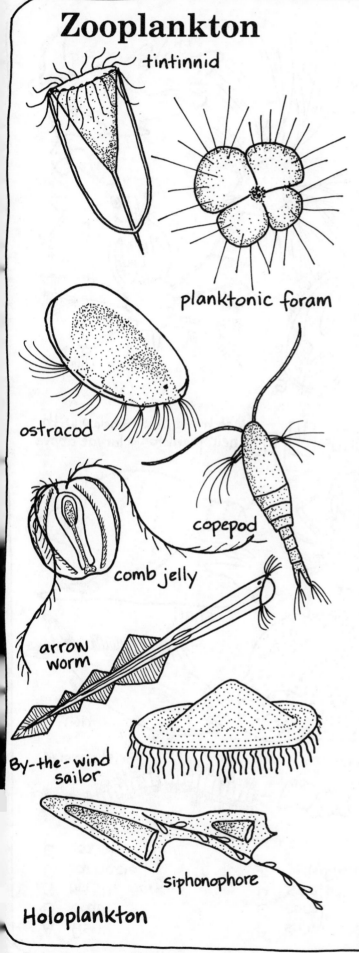

tintinnid

planktonic foram

ostracod

comb jelly

copepod

arrow worm

By-the-wind sailor

siphonophore

Holoplankton

Zooplankton are animal plankton. Species of zooplankton that spend their entire lives in a floating state are called *holoplankton*. Holoplankton are predominant in open ocean water (the pelagic zone). Temporary zooplankton, such as eggs and larvae, are known as *meroplankton* and are predominant in coastal waters (the neritic zone). Meroplankton exist as plankton for only a limited part of their development and later become either a part of the nekton or benthos. Being planktonic for part of their lives allows these temporary plankters to disperse into new areas.

Zooplankton can be further classified according to size. *Nannoplankton* (5/1000 mm to 60/1000 mm) include primarily protozoans, unicellular animals that often feed on phytoplankton and in turn are eaten by other zooplankters. *Microplankton* (60/1000 mm to one mm) are composed primarily of eggs and larvae, usually of invertebrates. *Macroplankton* (over one mm) often contain large numbers of copepods, along with amphipods, cumaceans and arrowworms. *Megaplankton* include mainly the large jellyfishes and their relatives, the Portuguese man-o-war and the By-the-wind sailor, which move at the mercy of the currents.

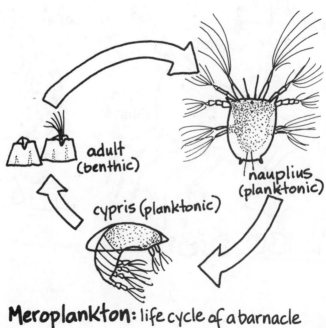

adult (benthic)

nauplius (planktonic)

cypris (planktonic)

Meroplankton: life cycle of a barnacle

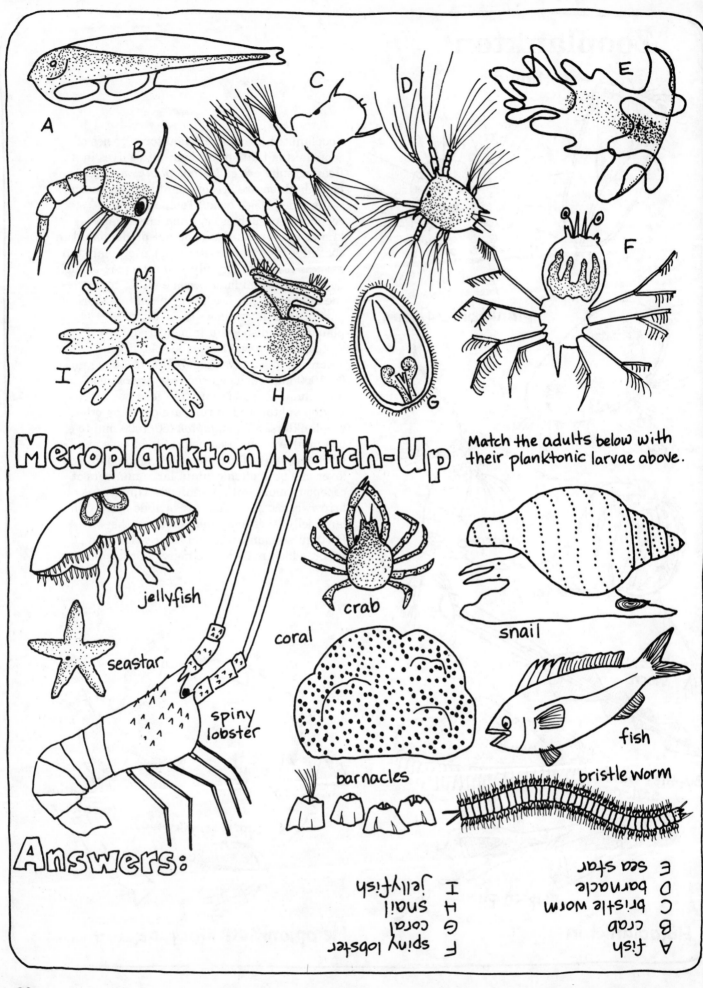

Meroplankton Match-Up

Match the adults below with their planktonic larvae above.

jellyfish

seastar

spiny lobster

coral

crab

barnacles

snail

fish

bristle worm

Answers:

I jellyfish
H snail
G coral
F spiny lobster

E sea star
D barnacle
C bristle worm
B crab
A fish

Nekton

All animals capable of swimming powerfully and purposefully comprise the *nekton*. Unlike the plankton, which move at the mercy of tides and currents, nekton are streamlined and move at speeds which make them independent of ocean currents. Nektonic creatures may begin life as plankton, however. Most fish have planktonic eggs, larvae, or both. Included in the nekton are the adult fish, marine mammals, and marine reptiles. Squid are the only invertebrates that are strong enough swimmers to be considered nekton.

Many nektonic animals are capable of long migrations, but very few can move at will throughout the entire ocean. Nonvisible barriers, such as gradual changes in temperature, salinity, and availability of nutrients effectively limit their range. Although a tarpon, for example, may be a strong enough swimmer to reach the Arctic from the Caribbean, it could not tolerate environmental changes such as the colder temperatures.

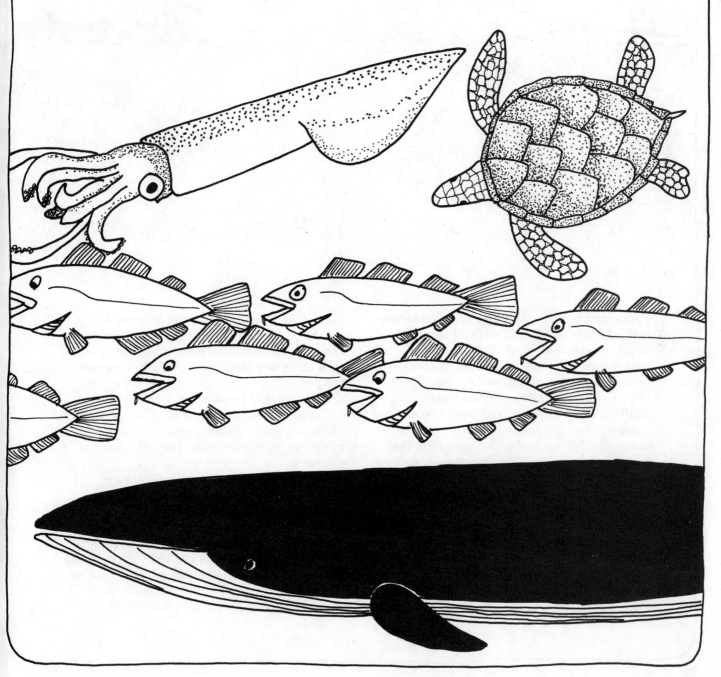

Benthos

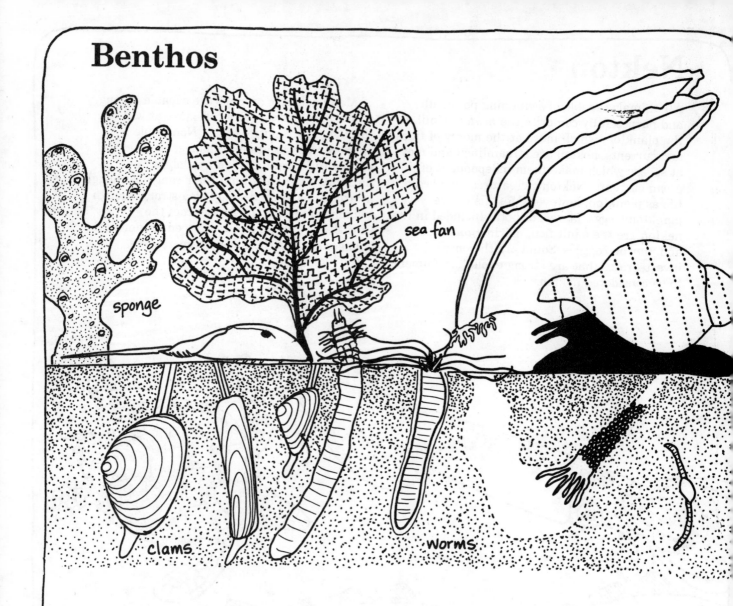

sponge

sea fan

clams

worms

Plants and animals living on or in the bottom are called the *benthos*. Nearly 16 percent of all living animal species are benthic. Organisms that live *on* the ocean bottom or on rocks, shells, seaweeds, pilings, etc. are called the *epiflora* (plants) and the *epifauna* (animals). Epiflora consist primarily of seaweeds and sea grasses. The epifauna need not necessarily be permanently attached to the substrate; they may just move over it. Sea anemones, sponges, corals, snails, and crabs are all epifauna. The epifauna include about four–fifths of all benthic animals.

Attached marine plants (epiflora) are found only in the intertidal and subtidal zones, the only benthic zones reached by light. The great majority of benthic animals also live in these zones (depths of less than 600 feet) because of the availability of food there. Most of the benthic environment is in perpetual darkness, and animals that live there must feed upon each other or upon food that falls down from the photic zone.

Animals that live buried *in* the bottom, such as clams and worms, are known as the *infauna*. There are about 30,000 species of infauna compared with more than 125,000 species of epifauna.

The larvae of benthic animals are an extremely important component of the meroplankton. It is estimated that 75 percent of the types of benthic invertebrates have a planktonic larval stage. Each animal may produce millions of eggs per year, yet only one or two will survive to adulthood; most will be eaten before settling to the bottom.

True/False Quiz

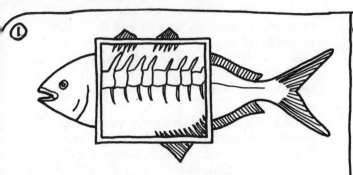

① 1. 60% of the animals on earth are vertebrates.

②

2. Some of the largest sharks eat plankton.

③ Phylum Echinodermata   Phylum Plankton

3. Plankton is a phylum.

④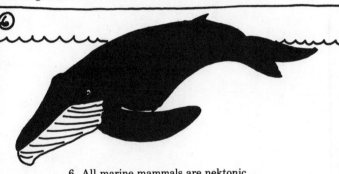

Who am I?

4. Organisms can have more than one scientific name.

⑤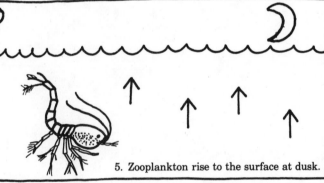

5. Zooplankton rise to the surface at dusk.

⑥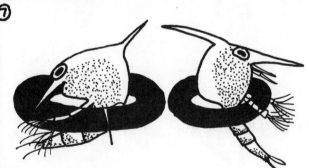

6. All marine mammals are nektonic.

⑦

7. Plankton cannot swim.

⑧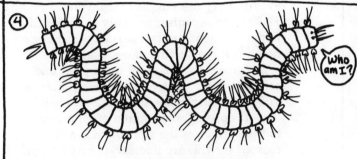

8. Mollusks evolved from coelenterates.

⑨

?

9. The Ocean sunfish is considered planktonic.

⑩ AUTOTROPH SHOP We sell plants

10. Plants are known as autotrophs.

23

1. *False* There are over a million described species of animals. Only 5 percent are vertebrates, that is, possess a backbone. All others are invertebrates; they have no backbone.
2. *True* The largest sharks, the Basking shark and the Whale shark, feed exclusively on plankton.
3. *False* Plankton are a group of organisms with a common lifestyle: They float at the mercy of the tides and currents. They are not a group of organisms with a similar evolutionary descent, as is a phylum.
4. *False* Each organism has only one scientific name, consisting of two parts: the generic name and the specific name. Many organisms, however, have more than one common name, such as *Crepidula fornicata*, known as the Slipper shell or Boat shell.
5. *True* It appears zooplankton often migrate to the surface as the sun goes down. The reasons for this are unclear. It may be that they are attracted to a particular light intensity, or perhaps they ascend to feed on phytoplankton.
6. *True* Marine mammals include whales, dolphins, and seals. None are considered benthic or planktonic. They are all capable swimmers.
7. *False* Many zooplankters swim quite well, especially in a small, calm dish of water. They are not strong enough, however, to swim against tides and currents.
8. *False* It appears mollusks and coelenterates followed dissimilar evolutionary paths.
9. *True* The Ocean sunfish, *Mola mola,* found all along the Atlantic coast, may grow to ten feet long, yet due to its feeble swimming ability, it is considered planktonic. It is a wanderer of the high seas and drifts at the mercy of the ocean currents.
10. *True* Plants produce their own food; "auto" means self and "troph" means food. Animals are known as heterotrophs.

3 Marine Plants

Flowering plants have true leaves, stems, roots, and flowers. Another term used to describe these plants is "vascular," meaning they contain vessels that conduct fluids up and down the plant. Flowering plants are not found in abundance in the ocean. Marine grasses and mangrove trees are among the very few vascular plants that grow in salt water.

The great bulk of marine plant life is made up of algae. Algae (singular: alga) are a diverse group of non-vascular plants. They do not have roots, stems, leaves, or flowers, although they may have similar-looking structures. Methods of sexual reproduction distinguish algae from all other plants.

Pelagic algae, that is, those living in the water column, are known as phytoplankton. The phytoplankton consist of free-floating unattached plants that are usually unicellular and microscopic, such as the diatoms and dinoflagellates (see Chapter Two).

Unlike the phytoplankton, most attached algae are multicellular and readily visible. The four major groups (divisions) of benthic algae are the blue-green algae, green algae, red algae, and brown algae. These groups are named for their dominant colors, but technically they are distinguished by the chemistry of their pigments.

Because of their photosynthetic activities, marine plants, especially algae, are important producers of organic matter in the ocean. Animal life depends on algae as its primary food source. In addition, marine plants are an important source of oxygen. During photosynthesis, they release vast quantities of oxygen into the water.

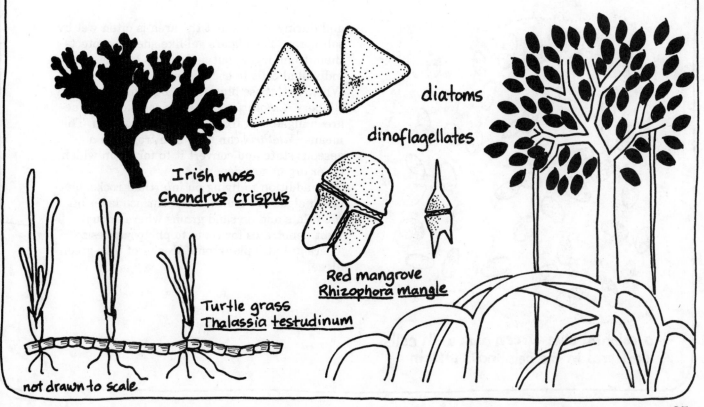

Irish moss
Chondrus crispus

diatoms

dinoflagellates

Red mangrove
Rhizophora mangle

Turtle grass
Thalassia testudinum

not drawn to scale

Blue-Green Algae

Fossil evidence indicates that blue-green algae (Division Cyanophyta or Cyanochloronta) have lived on earth for over two billion years, but their evolutionary origin remains obscure. Some consider the blue-greens to be bacteria because among living organisms, only bacteria and blue-green algae have cells with no membrane surrounding the nucleus. Blue-green algae, however, contain chlorophyll and they photosynthesize as other plants do, liberating oxygen in the process. In fact, blue-green algae are often thought to be responsible for the production of atmospheric oxygen early in the earth's history.

Blue-green algae are found almost anywhere light and water are present: in the air, on the ground, and in fresh and salt water. Individual blue-green algal plants are microscopic, but in marine species they usually cluster together in colonies covered with a gel-like sheath.

On temperate shores, blue-green algae are most commonly found as dark films on intertidal rocks or as tiny epiphytes on other algae. One variety, *Calothrix*, grows in thin mats that form a conspicuous black zone on rocky shores near the high tide mark. Seawater reaches these barren rocks only at higher tides

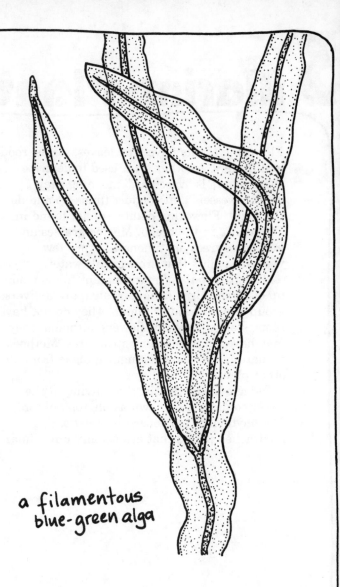

a filamentous blue-green alga

and during storms, but the area is often wet by salt spray. The algae's gel-like sheath protects them from drying out and serves to glue individual cells to each other and to the rocks. When wet, these blue-green algal mats are very slippery. *Calothrix*, as well as some other blue-green algae, can fix atmospheric nitrogen. This means *Calothrix* can take in nitrogen in a gaseous state and convert it to forms on which other organisms depend.

In addition to living on intertidal rocks, blue-green algae are found in the surface layer of mud flats and on sand grains where enough light penetrates for them to photosynthesize. There are also planktonic species of blue-green algae.

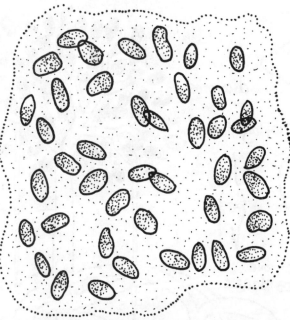

a colonial blue-green alga with cells scattered in the gelatinous sheath

Green Algae

In fresh water, the most conspicuous algae are the greens (Division Chlorophycophyta). In the sea, green algae are far less common. In fact, only 13 percent of green algal species are marine. Many benthic green algae (green seaweeds) are large enough to be seen without a microscope. Some species can grow to several feet in length. On the shore, they are easily recognized as green algae simply by their grassy green color. The pigments in green algae are similar to those of flowering plants, thus the same green color.

Green seaweeds have a variety of shapes, from filaments and sheets to cylinders and spheres. They all need plenty of light, and are most common at higher intertidal zones or subtidally in shallow water. Cells of green algae may have more than one nucleus.

Almost all seaweeds (green, red and brown) are attached plants, fixed to the substrate by a structure known as a holdfast. A holdfast may resemble the root system of vascular plants, but it has an entirely different function. It serves only to anchor the plant, not to absorb water and nutrients, as roots do. Although algae come in many shapes and forms, generally there is a stem-like structure called the stipe and a leaf-like structure called the blade.

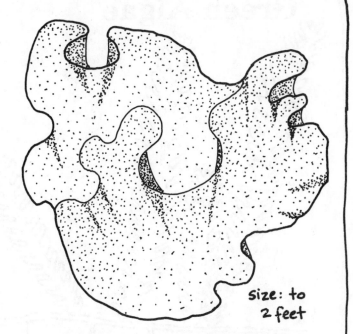

size: to 2 feet

Sea lettuce _Ulva lactuca_

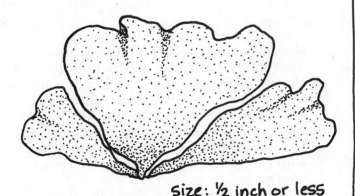

size: ½ inch or less

Prasiola stipata

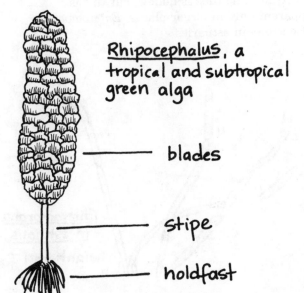

Rhipocephalus, a tropical and subtropical green alga

—— blades

—— stipe

—— holdfast

Sea lettuce, _Ulva lactuca,_ grows near and below the low tide mark from Maine to Florida. It is a bright green alga made up of a double layer of cells in flat or ruffled sheets. The blade is translucent and has the consistency of wet wax paper. Its holdfast is inconspicuous. Sea lettuce is an edible seaweed. It can be added to salads or can be dried and powdered to use as a seasoning.

Prasiola stipata grows in dirty, flaky green patches found in the spray and upper intertidal zones from Newfoundland to Cape Cod. It usually grows in areas with lots of bird droppings, especially around gull rookeries.

Green Algae

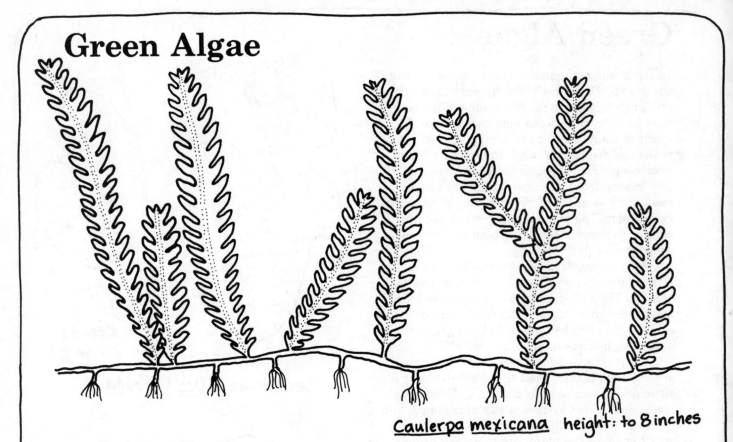

Caulerpa mexicana height: to 8 inches

More than twenty species of *Caulerpa* grow along the coast of Florida and the Gulf of Mexico. Caulerpas have long stalks that run along the surface of the substrate. Blades and holdfasts branch off from these stalks. The variety among the different species of *Caulerpa* is amazing. Some have feathery blades, while others are disc-like, but the plants are always a vivid green.

Valonia ventricosa is a unique alga. This green seaweed, known as Sea bottles, grows in shallow water in southern Florida and the Caribbean. It is a large, thin-walled sac resembling a clear green marble. This sac is actually a single cell with many nuclei, and is among the largest single cells known. Growing under the sac are microscopic cells that form tiny root-like holdfasts. *Valonia* grows attached to hard objects in areas of clear water and good currents.

Although *Enteromorpha intestinalis* is found mostly in the intertidal zone from the Arctic to South Carolina, similar species are found as far south as Florida. Species of *Enteromorpha* are thin tubular green algae that arise from a small disc at their base, where they attach to the substrate. The tube is hollow, but this is apparent only in larger plants. *Enteromorpha* also grows in estuaries.

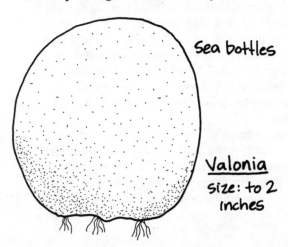

Sea bottles

Valonia
size: to 2 inches

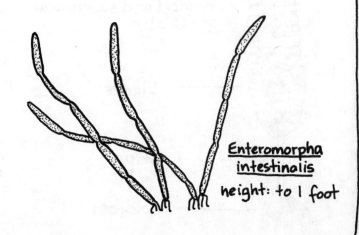

Enteromorpha intestinalis
height: to 1 foot

Red Algae

There are more species of red seaweed (Division Rhodophycophyta) than of brown and green seaweeds combined. The vast majority of red algae are marine; less than 2 percent live in fresh water. Generally, red algae are able to grow under low light conditions because of their accessory pigments which are very effective at capturing light. Consequently, some species of red algae are found at depths as great as 600 feet. Red seaweeds often have beautiful forms. They may be tiny epiphytes on other algae, or large fleshy plants attached to rocks.

Red seaweeds grow in all latitudes, but they predominate in tropical and subtropical regions. Brown and green algae grow mostly in polar, sub-polar, and temperate regions.

Unlike the green seaweeds, the color of red algae is not uniform. Several species look yellow or brown, the red color being masked by other pigments. The most definite difference between red algae and other algae is that unlike brown and green algae, red algae do not produce any flagellated reproductive bodies.

Many red seaweeds are commercially important, especially as food. Do you enjoy chewing on salty rubber bands? If so, Dulse, *Rhodymenia palmata,* is for you. Dulse is an edible seaweed that grows from Long Island northward, from mid-intertidal levels to deep water. It can be eaten raw (rubbery) or dried (better) or used as seasoning in salads and soups. Live Dulse is deep red and nearly opaque.

Purple laver, or Nori, *Porphyra,* is another edible red algae. This seaweed is transparent and paper thin, and except for its color, strongly resembles Sea lettuce. Nori is quite tasty in sandwiches. It can also be seasoned and eaten with rice. It grows from Florida to Maine, on rocks, pilings, shells, and other algae.

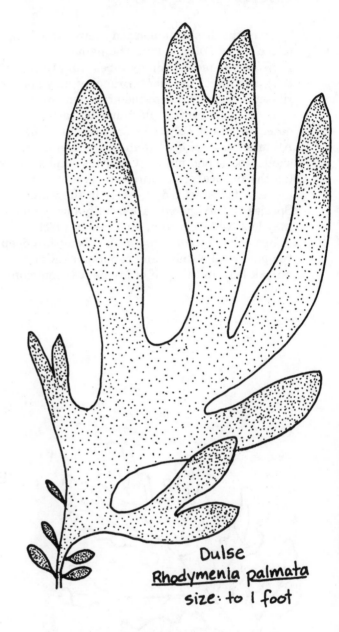

Dulse
Rhodymenia palmata
size: to 1 foot

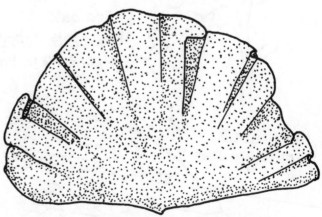

Nori or Purple laver   Porphyra
size: to 1 foot

Red Algae

Some of the most common red seaweeds belong to the genus *Polysiphonia*. The name *Polysiphonia* means "many tubes" and refers to the structure of the plant. Branches may appear ribbed or tubular due to the arrangement of cells around a central filament. At least two dozen species of *Polysiphonia* grow along the Atlantic coast. Many live in tidepools and on shells, rocks, and pilings. One lives epiphytically (on the surface) on Knotted wrack, a brown algae. A few species drift freely in the water. Species of *Polysiphonia* are usually bushy and may look yellow, brown, pink, red or black.

Sea oak, *Phycodrys rubens,* is a beautiful deep red alga with a shape and venation similar to an oak leaf. In the Gulf of Maine it is commonly found as an epiphyte on coarse seaweeds such as kelp. Sea oak is a good species to use in seaweed pressing.

Eucheuma is a genus of tropical algae with a very high carrageenan content. *Eucheuma* is very rubbery and feels like it would be sold at a dime store toy counter. Even its color, which is usually yellow or brownish red, makes it look like plastic. *Eucheuma* has many knobby branchlets and grows in clumps or mounds.

Gracilaria is a bushy red alga found along the entire coast. Along the southeast coast it is harvested as an important source of agar, which is used for bacterial cultivation in hospital laboratories, for cosmetics, hand lotions, shoe polish, and photographic processes.

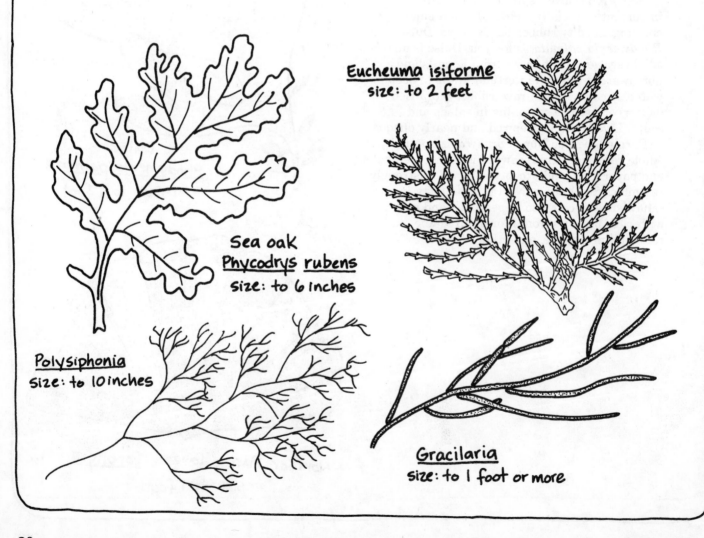

Sea oak
Phycodrys rubens
size: to 6 inches

Eucheuma isiforme
size: to 2 feet

Polysiphonia
size: to 10 inches

Gracilaria
size: to 1 foot or more

Pressing Seaweed

Few people think of seaweeds as beautiful. Along the shore, out of water, seaweeds are often seen in sticky globs or in sun-dried heaps. But by isolating a particular specimen of seaweed through floating and pressing it, the delicate structures and vivid colors of the plant become evident. Shown on this page is a red seaweed, *Hypnea musciformis*, which grows all along the Atlantic coast.

Directions:

1. Take a piece of seaweed from the shore.
2. Wash it with sea water.
3. Fill a flat pan with sea water. Float a bit of the seaweed in it until it opens up fully (figure A).
4. Slide a sheet of heavy or stiff paper under the seaweed in the water. Index cards (five inches by seven inches) work well (figure B).
5. Tip the paper to drain off the water, keeping the seaweed on top.
6. Place wet paper with the seaweed on newspapers.
7. Cover with waxed paper, then with more newspapers.
8. Put a heavy object, such as a book, on top (figure C).
9. The next day, replace both the newspaper and waxed paper with dry sheets to prevent the seaweed from molding.
10. Repeat step nine the next day. After four days the seaweed should be dry. Some seaweeds have a natural mucilaginous adhesive which will keep them stuck to the paper; others may need a bit of glue.

If you have brushed your teeth or eaten ice cream today, you are already familiar with the gelatinous properties of Irish moss, *Chondrus crispus*, a dark red seaweed that grows intertidally from Newfoundland to New York. Carrageenin, a gelatinous colloid extracted from Irish moss, is used as a stabilizer and suspending agent in hundreds of products from toothpastes and ice cream to soups and medicines. Irish moss is harvested in New England and Nova Scotia for this purpose. *Eucheuma*, a red seaweed found on the Florida coast, also contains lots of carrageenin.

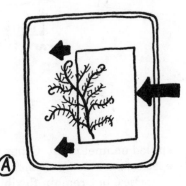
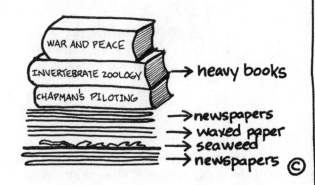

Blanc Mange Pudding

Blancmange pudding is a tasty way to use Irish moss or *Eucheuma*:

1. Add one cup of washed Irish moss or *Eucheuma* to one quart of milk.
2. Cook in a double boiler for half an hour.
3. Strain the milk to remove the seaweed.
4. Add vanilla, sugar, and honey, to taste.
5. Refrigerate until firm.

Calcareous Algae

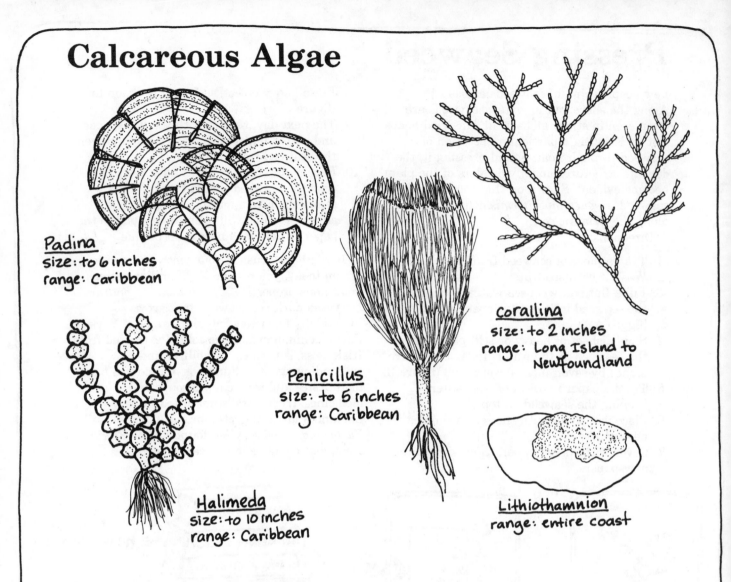

Padina
size: to 6 inches
range: Caribbean

Halimeda
size: to 10 inches
range: Caribbean

Penicillus
size: to 5 inches
range: Caribbean

Corallina
size: to 2 inches
range: Long Island to Newfoundland

Lithiothamnion
range: entire coast

The seaweeds shown on this page are just a few of the many species of algae that deposit limestone (calcium carbonate) within their tissues. If enough limestone accumulates, the algae become very hard. When calcareous algae die, their living tissue decomposes but their white limestone "skeleton" remains. Red algae with heavy accumulations of limestone lose all superficial resemblance to plants.

One family of red algae, Corallinaceae, is the most prevalent of all calcareous marine algae. Named for their superficial resemblance to hard coral, it was not until the mid 1800's that coralline algae were recognized as plants. Corallinaceae contains many species of red algae, some of which grow as stony crusts, such as *Lithothamnion*, the pinkish red crust found on rocks in New England tidepools as well as on coral reefs in the Florida Keys. It looks like spilled enamel paint. Other Corallinaceae species, such as *Corallina*, grow erect, with branches, but remain flexible because the joints between their segments are not calcified.

Although the majority of calcareous algae are red, there are a few green and even fewer brown calcareous species, most of which are tropical. *Padina* is a brown algae that often accumulates calcium carbonate over its fan-shaped blade. *Penicillus*, a green algae, also becomes calcified. Neither *Padina* or *Penicillus*, however, becomes crunchy, just stiff. *Halimeda*, like *Corallina*, has crunchy segments alternating with flexible joints.

Many red and green algae (especially *Halimeda*) build up calcareous sediments as they die, producing much of the sediment in southern Florida and in tropical lagoons, beaches, and reefs.

Brown Algae

Most of our familiar large seaweeds, including rockweeds, kelp, and *Sargassum*, are brown algae (Division Phaeophycophyta). Brown algae are often an olive-brown color. They grow along rocky shores in all oceans, but often predominate in temperate and cold seas. They are usually found near the shore and in water not deeper than sixty feet. Nearly all species of brown algae are marine. A few brown seaweeds grow as crusts or filaments, but most are medium to large leafy plants.

Dictyota is a delicate but crisp brown algae. Species of *Dictyota* are restricted to warmer climates, from Florida to the tropics. They grow as small bushy plants attached to rocks and

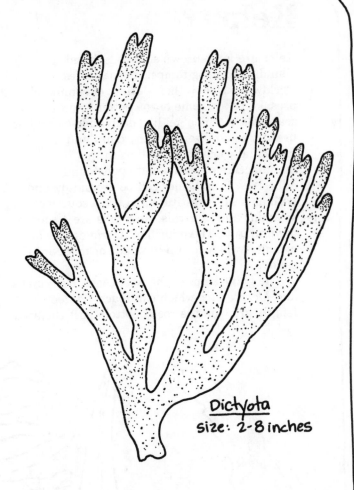

Dictyota
size: 2-8 inches

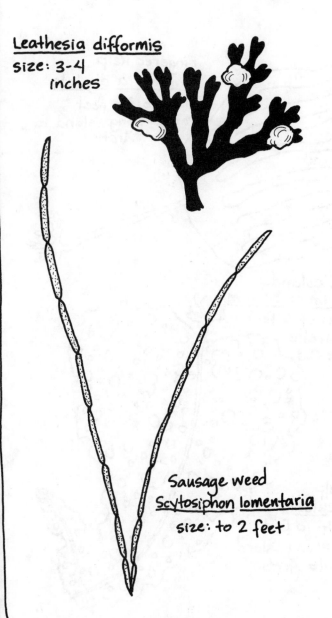

Leathesia difformis
size: 3-4 inches

Sausage weed
Scytosiphon lomentaria
size: to 2 feet

other hard objects in fairly shallow water. The symmetrical, regular branching pattern of *Dictyota* makes it a very attractive plant.

Sea potato, *Leathesia difformis*, grows intertidally and subtidally from Newfoundland to North Carolina. This hollow, lumpy, yellow-brown algae grows epiphytically in sac-like clumps on large seaweeds and on rocks. As illustrated here it is shown growing on Irish moss, *Chondrus crispus*, which it may eventually choke out. Sea potato is rubbery and, except for the fact that it is not porous, looks a lot like a sponge.

Scytosiphon lomentaria, commonly called Sausage weed, is appropriately named. Mature plants grow twisted and constricted at intervals, like a chain of sausages. Sausage weed grows intertidally on rocks and jetties from Canada to Florida, reaching its largest size (two feet) in cold northern waters. Large aggregations of this algae harbor communities of small fishes and invertebrates.

Kelp

Kelps are large brown seaweeds found abundantly along temperate coastlines worldwide. Due to their variety of commercial uses, kelps are some of the better known seaweeds. Algin, a gel-like derivative of kelp, is used as a stabilizer in many dairy and medicinal products as well as in the manufacture of paints, auto polish and adhesives. Kelp blades can be ground up and used as a salt substitute and are a source of vitamins and minerals. Kelp pills are a source of vitamin C. The midrib of *Alaria esculenta* is delicious when cut up in salads or made into candy.

The giant kelps of the west coast, *Nereocyctis* and *Macrocystis,* which can reach 120 feet in length, grow on the rocky bottom past the area where the waves break. They are harvested mechanically by large mowing machines mounted on barges. A cutting bar on the mower is set to cut the blades four feet below the water surface, and the cut blade tips are brought aboard the barge on a conveyor. Blades grow back rapidly.

Kelps of the Atlantic coast are smaller, reaching ten feet in length, and grow primarily below the low tide line. Several species are illustrated here.

Unlike the stems of vascular plants, the stipes of algae usually do not serve a conductive function. But some species of kelp have cells in their stipe that form sieve tubes which transport food up and down within the plant.

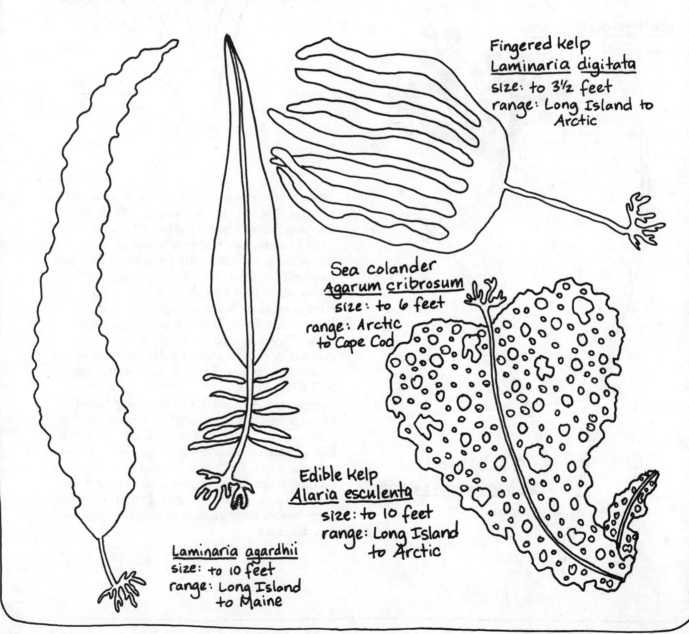

Fingered kelp
Laminaria digitata
size: to 3½ feet
range: Long Island to Arctic

Sea colander
Agarum cribrosum
size: to 6 feet
range: Arctic to Cape Cod

Edible kelp
Alaria esculenta
size: to 10 feet
range: Long Island to Arctic

Laminaria agardhii
size: to 10 feet
range: Long Island to Maine

Rockweeds

Those large, slippery brown seaweeds that grow in dense beds along the Atlantic coast are the rockweeds. Bladder wrack, *Fucus vesiculosus*, and Knotted wrack, *Ascophyllum nodosum*, are the most common rockweeds. They account for most of the seaweed in the rocky intertidal zone of New England.

Air bladders in the rockweeds help buoy up the plants when the tide comes in, ensuring that the rockweeds get enough light for photosynthesis. Alginates, gel-like substances in the cell walls of rockweeds, provide protection from desiccation while lending flexibility and strength. This allows the algae to tolerate the crashing waves of the intertidal zone. Communities of plants and animals thrive on rocks below the lush protection of the rockweeds and on the rockweed blades themselves.

Bladder wrack, found from the Arctic to North Carolina in exposed areas, has a midrib and paired air bladders. Warty reproductive structures grow at the tips.

Knotted wrack grows intertidally from Long Island Sound to the Arctic. Sometimes pieces of Knotted wrack break off, float away, and wash ashore far south of their usual range. Knotted wrack has no midrib, but does have air bladders. Flagellated gametes (sex cells) are released into the water from the reproductive structures and fuse together to start a new rockweed plant.

Both rockweeds can be used for poultry meal, fertilizer and garden mulch.

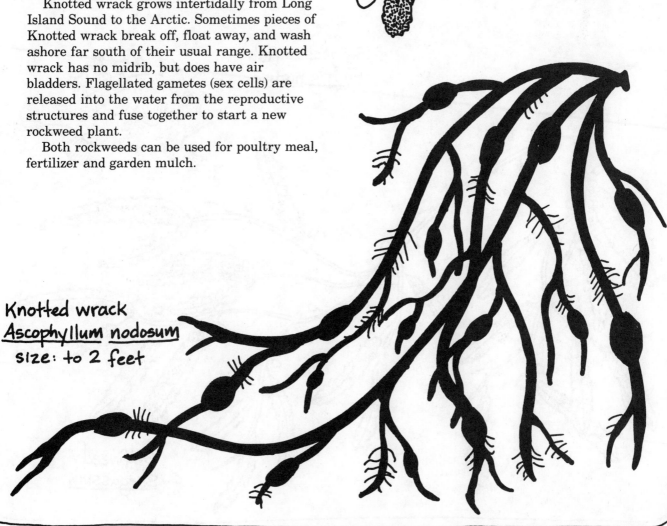

Bladder wrack
Fucus vesiculosus
Size: to 3 feet

Knotted wrack
Ascophyllum nodosum
Size: to 2 feet

Sargassum

One of the few brown seaweeds that grows abundantly in the tropics is *Sargassum,* commonly called Gulfweed. *Sargassum* has air bladders on short stalks and is a golden brown color. Fifteen different species grow along the Atlantic coast from Maine to Florida, although three are especially common. *Sargassum filipendula* is benthic (attached) and is found from Cape Cod south; it grows very thickly just south of Cape Hatteras, North Carolina.

Sargassum natans and *Sargassum fluitans* are pelagic, that is, free-floating. Often they are found washed ashore, particularly south of Cape Cod. These pelagic species are the principal seaweed of the Sargasso Sea. Located south of Bermuda, where surface currents converge in an area of calm water, the Sargasso Sea covers an area two-thirds the size of the United States although the algal layer is never thick enough to hinder navigation. Seven million tons of live *Sargassum* float here unattached. The plants reproduce through the breaking off of pieces, a form of asexual reproduction.

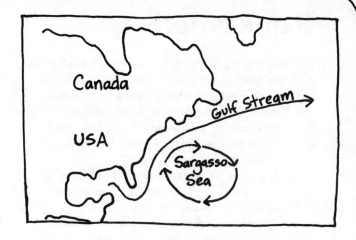

The drifting masses of seaweed in the Sargasso Sea, along with clumps that float away and wash ashore, support a remarkable community of animals found nowhere else. Hydroids, bryozoans, worms, shrimp, crabs and fish are just a few of the creatures that live in *Sargassum.* Many of these animals exhibit amazing mimicry, having adaptations which imitate the color and form of *Sargassum.*

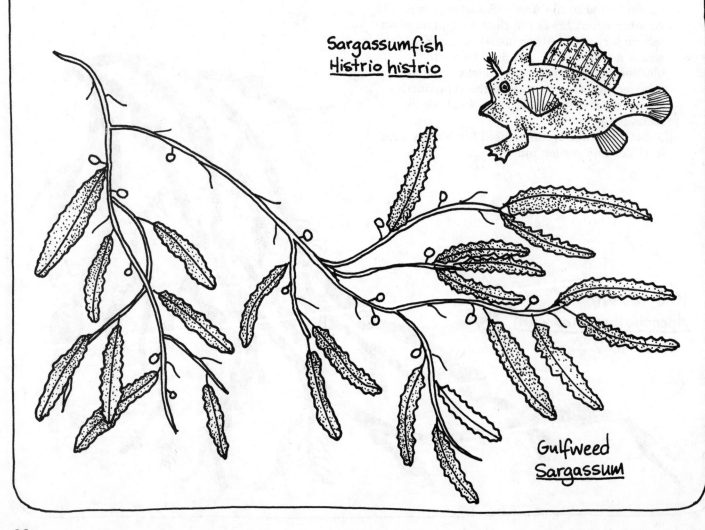

Sargassumfish
Histrio histrio

Gulfweed
Sargassum

Marine Grasses

Approximately seven species of marine grasses grow along the coast of North America. Unlike algae, marine grasses are true flowering plants. They have roots, stems, leaves and flowers. Eelgrass, *Zostera marina*, and Turtle grass, *Thalassia testidinum*, are the predominant species on the Atlantic and Gulf coasts.

Marine grasses grow wholly submerged in salt water. Strong roots anchor the grasses into the sediment. Flowering and pollination occur underwater. Pollen from the grasses drifts in water currents and attaches to the stigma of the flower with which it collides. Vast beds of sea grasses form important nursery grounds for many invertebrates and fishes.

Eelgrass grows along the Atlantic coast from southern Greenland to South Carolina in sandy flats and sheltered inlets. Its roots penetrate the mud and mat together tightly, stabilizing the substrate. Usually Eelgrass grows just below the low tide line, although if the current is not over-powering and the water is clear enough to allow maximum light penetration, Eelgrass can be found in 100 feet of water.

Aside from Canada geese, Brants, a few ducks, and some invertebrates, few animals eat living Eelgrass. Most eat the grass as it decays. The blades also provide attachment sites and shelter for many plants and animals. In 1930–1931 an epidemic destroyed 90 percent of the Eelgrass beds on the Atlantic coast, and beds are still recovering.

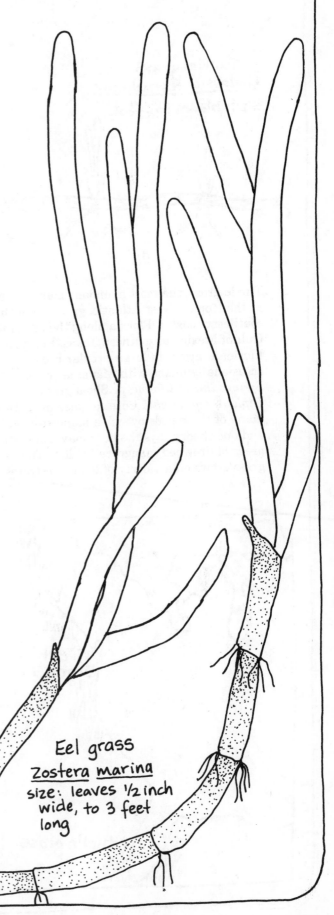

Eel grass
Zostera marina
size: leaves ½ inch wide, to 3 feet long

Marine Grasses

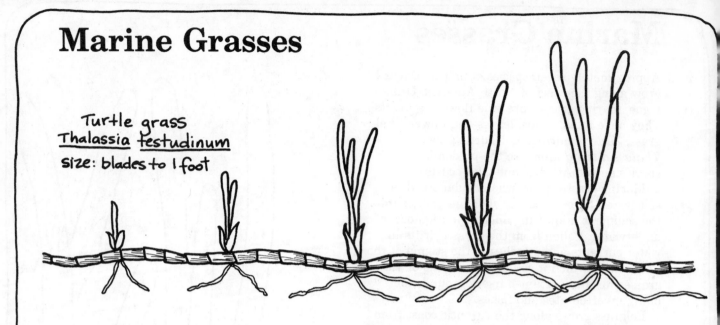

Turtle grass
Thalassia testudinum
size: blades to 1 foot

Turtle grass, the most abundant marine grass in the tropical west Atlantic, grows from the southeast coast of Florida along the arc of the Gulf of Mexico to southern Texas. In southern Florida it grows in large circular beds, occasionally mixed with Manatee grass, *Syringodium filiforme,* or Shoal grass, *Diplanthera wrighti.* Core samples indicate that many of these beds overlie a mangrove peat, and one theory says the beds may dominate areas of drowned mangrove habitat. Turtle grass grows on a variety of loose substrates such as sand, mud, and broken shells. It favors areas of clear calm water.

Turtle grass has rhizomes (underground stems) that penetrate three to six inches below the substrate. Erect shoots develop from these stems, consisting of a short shoot bearing four or five leaves with a sheath at the base.

As with Eelgrass, more animals eat the decomposing Turtle grass blades than eat the live blades. Broken leaf blades wash up on the shore and are eaten by detritivores such as amphipods and snails.

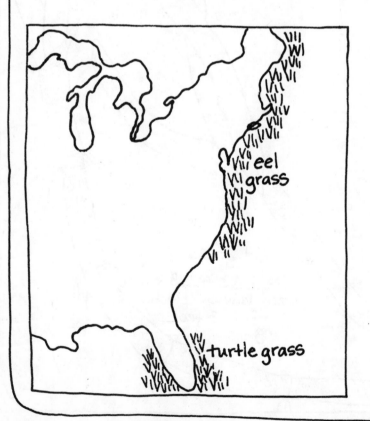

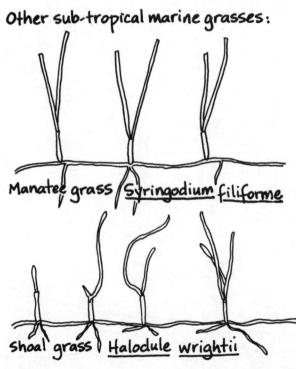

Other sub-tropical marine grasses:

Manatee grass _Syringodium filiforme_

Shoal grass _Halodule wrightii_

Mangroves

Aside from marine grasses, which grow wholly submerged in salt water, the other flowering marine plants are mangroves, which grow partially submerged along shallow tropical and subtropical seashores around the world. In the United States, mangrove trees grow along the south coast of Florida.

Of the three types of mangroves in Florida, the Red mangrove, *Rhizophora mangle*, generally grows closest to the water's edge, where its roots are flooded at high tide. Its tall anchoring prop roots hold the tree firmly in the unstable ground despite the uprooting pressure of wind and waves. Thus, mangroves stabilize the substrate and hold the sediments.

Mangroves are terrestrial plants with special adaptations allowing them to utilize sea water. The roots of the Red mangrove filter some of the salt out of the water, while some salts accumulate in the leaves, which eventually fall off the tree. Fallen mangrove leaves are an important source of food and detritus in the typically nutrient-poor tropic and subtropic seas.

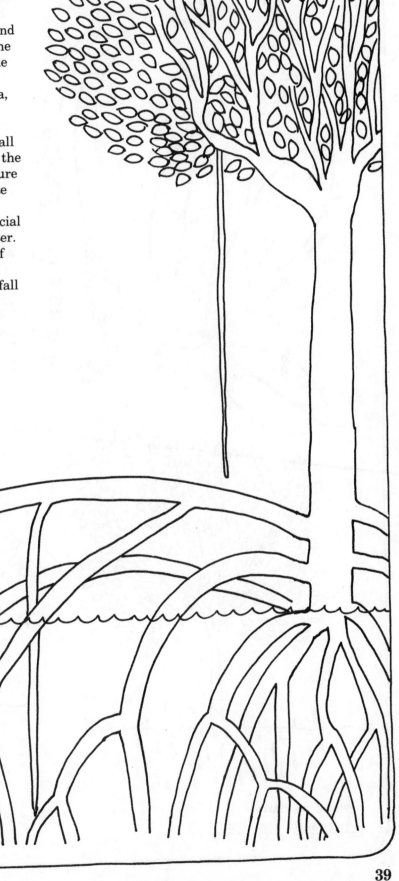

Red mangrove
Rhizophora mangle

Mangroves

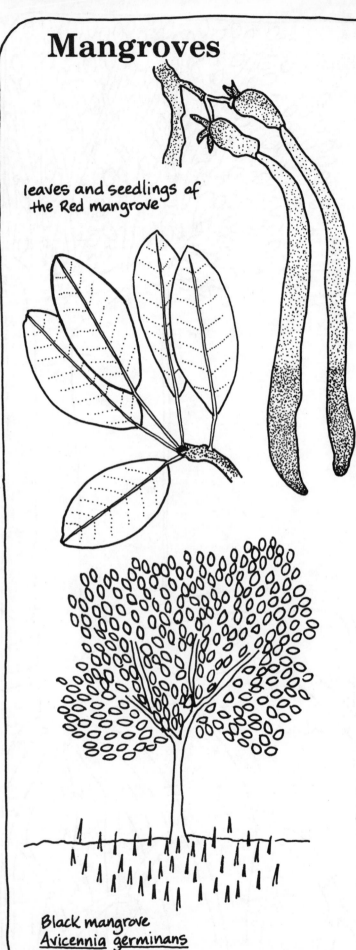

leaves and seedlings of
the Red mangrove

Red mangrove leaves are leathery, with sunken stomata. Both these adaptations help to prevent the loss of fresh water due to evaporation from the leaves.

While still on the tree, seeds of Red mangroves germinate into partially-rooted cigar-shaped seedlings called propagules. When the propagules drop, they may root in the mud under the tree or float upright until taking root in a suitable spot. The propagules may float for a year and still remain viable. Red mangroves have the ability to form a pioneer community, that is, to be the first plants to colonize a particular area.

Black mangroves, *Avecinnia nitidia*, have breathing roots called pneumatophores, which stick straight up out of the mud, providing aeration for the underground roots. Excess salt is excreted through the Black mangrove's leaves, which often have visible salt crystals on the surface.

White mangroves, *Laguncularia racemosa*, usually live among, or further inland from, the Black mangroves. They may have pneumatophores. Two salt pores on the leaf stem excrete excess salt.

Black mangrove
Avicennia germinans

White mangrove
Laguncularia racemosa

Marine Plant True/False Quiz

1. Algae produce seeds.

2. The Red Sea got its name from blue-green algae.

3. The minute Tardigrade, or water bear, is about the only creature that eats blue-green algae.

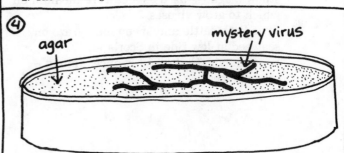

4. Green algae yields the substance agar, which is used in biomedical studies.

5. Along the coast of France, sheep wander to the shore to feed on seaweeds.

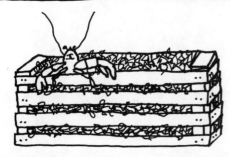

6. Algae absorb nutrients, carbon dioxide, and water through all their surfaces.

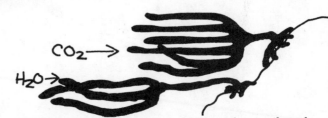

7. Calcareous algae are often used for packing baitworms and lobsters.

8. Turtles do not eat Turtle grass.

9. In preparation for hurricanes, people often tie their boats to mangroves.

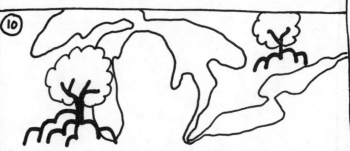

10. Red mangroves can grow in fresh water.

41

1. *False* Only flowering plants produce seeds. Algae reproduce in other ways.
2. *True Trichodesmium* is a blue-green alga rich in reddish pigment. When very abundant, *Trichodesmium* produces a slick which discolors the water. The Red Sea was named because of this periodic discoloration.
3. *False* Many other animals, including protozoa and intertidal snails, feed on blue-green algae.
4. *False* Red algae yields agar. In biomedical studies, agar is used as a culture medium on which to grow viruses.
5. *True* Both cattle and sheep may graze on seaweed at low tide in Scotland, Scandinavia, Chile, and France, among other places.
6. *True* For this reason, the need for specialized structures such as roots, stems, and leaves is eliminated.
7. *False* Knotted wrack, *Ascophyllum nodosum*, is often used to pack baitworms and shellfish.
8. *False* Green turtles eat Turtle grass. Manatees eat Manatee grass.
9. *True* The prop roots of Red mangroves are anchored so firmly in the substrate that they are more hurricane resistant then man-made docks or moorings.
10. *True* Although they cannot grow in the Great Lakes, Red mangroves thrive in tropical fresh water. But they are often out-competed by other plants in fresh water habitats, and so relegated to salt water shores, where there are fewer plants with which to compete.

4 Marine Protozoa

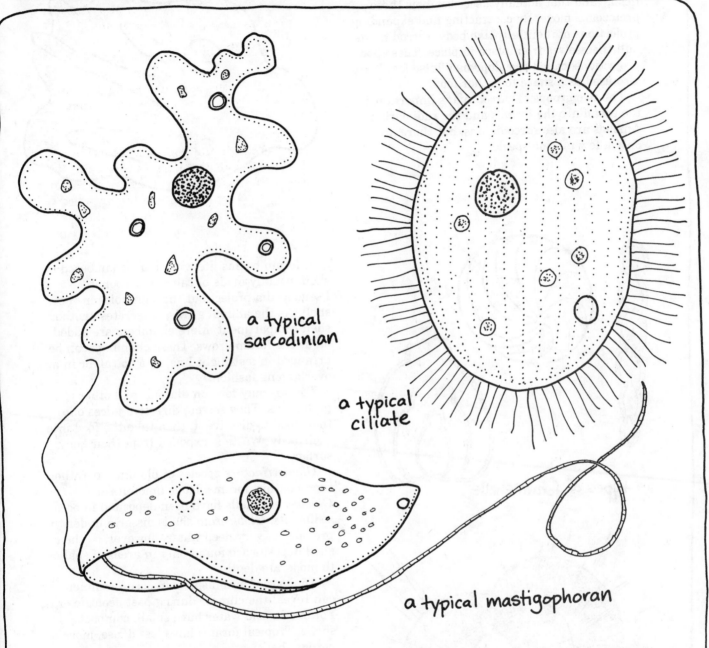

a typical sarcodinian

a typical ciliate

a typical mastigophoran

Protozoans are one-celled organisms. They are usually microscopic, and most live in water. Microscopic examination of a drop of sea water or a scraping from a rock or piling will almost always reveal a protozoan.

Whether single celled or multicellular, all organisms must carry on activities essential to survival: obtaining food, respiring and eliminating waste products. Although a protozoan is only one cell, it *is* a complete organism and carries out all these functions through organelles within the cell. Three groups

(phyla) of protozoans common in the ocean are the sarcodinians, the ciliates, and the flagellates.

Some protozoans are common in the plankton but others are benthic. Marine sediments contain many benthic protozoans and tiny multicellular animals collectively known as the meiofauna. These minute creatures climb, crawl, scuttle, and swim between sand grains, feeding on bacteria, other protozoans, diatoms, and tiny bits of organic matter.

Sarcodina

The word sarcodina literally means "creeping flesh," and that actually describes how these protozoans move. By contracting and expanding projections of their jelly-like body, sarcodinians pull themselves from place to place. These body projections, called pseudopodia ("false feet") are also used as a snare to gather food.

The two groups of sarcodinians important in the ocean are the forams and the radiolarians. Amebas are also sarcodinians, but there are only a few marine species.

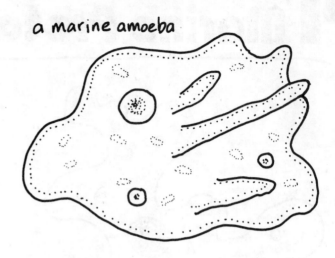

a marine amoeba

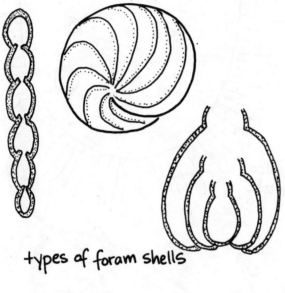

types of foram shells

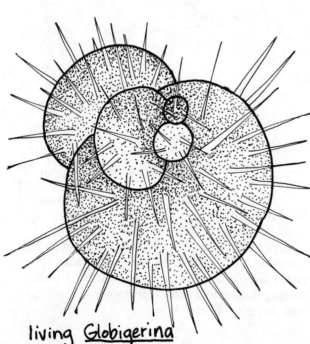

living Globigerina

A foram begins life with a one-chambered shell, usually made of calcium carbonate. Pseudopodia project out through holes in the shell. As the animal grows, it secretes another shell compartment. More chambers are added on as the foram grows. These chambers can be arranged in a straight line, in a spiral, or in an overlapping fashion.

Forams may feed on diatoms and other protozoans. They secrete digestive juices onto their food to dissolve it, then take the food into their body. Waste is expelled from their body surface.

Globigerina are species of planktonic, rather then benthic, forams. Huge amounts of *Globigerina* shells have been deposited in ocean sediments. *Globigerina* shells dissolve under the pressure of very deep water, however, so their shell deposits are found only in areas of shallow to moderate depth.

Analysis of foram shells in ocean sediments can reveal the climate during past geologic eras. Forams in cold water have small, compact shells. Tropical forams have less dense, more porous shells.

Sarcodina

Most species of forams are benthic rather than planktonic. Although benthic forams are usually very tiny, a few species in south Florida and the Gulf of Mexico are large enough to be seen with the naked eye. *Archaias angulatus*, the Turtle grass foram, for example, is very common on Turtle grass blades. Its shell is white and coiled. Another example, *Archaias compressus*, the Button foram, is thin and flat. It is nearly circular and its shell is a tightly coiled series of chambers. Shells of both these forams are a component of beach sand in the Caribbean.

Radiolarians, the other group of Sarcodinians, are entirely marine and mostly planktonic. A perforated outer skeleton of silica covers their spherical body. Pseudopodia extend through the holes as long, sticky filaments. Their skeleton may have spines or it may be more of a latticework. Some radiolarians are not spherical at all, but instead have skeletons in very bizarre patterns.

Unlike the calcium carbonate skeleton of forams, the siliceous skeleton of radiolarians does not dissolve at great depths. It is found in deep ocean sediments, whereas *Globigerina* skeletons are found only at shallow depths.

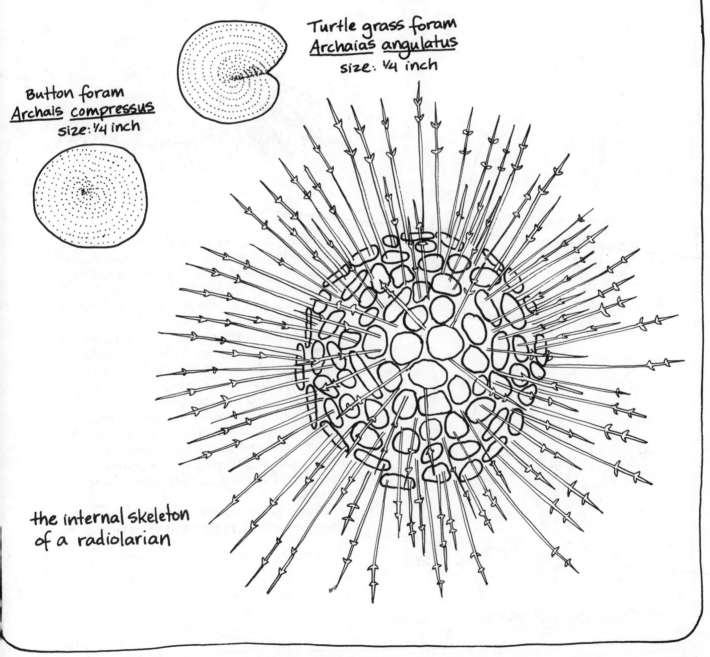

Turtle grass foram
Archaias angulatus
size: 1/4 inch

Button foram
Archais compressus
size: 1/4 inch

the internal skeleton of a radiolarian

Ciliates and Mastigophorans

Another group of important marine protozoans are the ciliates. As their name implies, they are covered with hair-like cilia. The cilia are used in eating, locomotion, and respiration. Most ciliates are solitary and free-swimming, but some are attached and colonial. Ciliates are extremely common among sand grains, where they eat plant cells and bacteria. In turn, ciliates serve as food for many larger animals.

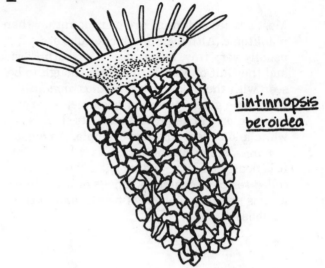

Tintinnopsis beroidea

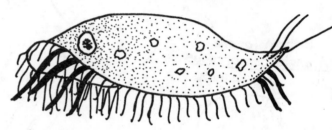

another ciliate

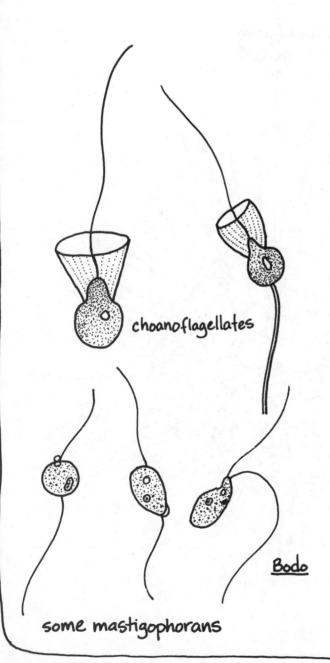

choanoflagellates

Bodo

some mastigophorans

Tintinnids, known as "bell animals," are planktonic ciliates common in the open ocean. A goblet-shaped hard shell of protein or foreign particles cemented together surrounds the tintinnid. A ring of cilia surrounding the mouth is used for locomotion and catching food.

Protozoans in the phylum Mastigophora are known as the flagellates. They are propelled by the beating action of their whip-like flagella. Choanoflagellates are a group of flagellates that are colonial and live attached to the substrate. Examples are *Monosiga ovata* and *Codonosiga botrytis*. Some have a stalk; others are attached directly. A cylindrical collar around the base of their single flagellum filters fine particles from the water. Food particles are engulfed and digested.

Bodo is an example of a free-living marine flagellate with two flagella. It feeds on bacteria.

Protozoan True/False Quiz

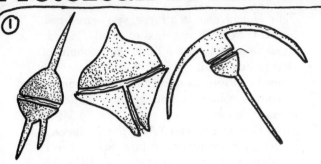

1. Dinoflagellates are often considered protozoans.

2. Protozoans do not live in the soil.

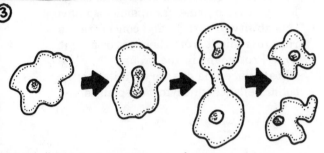

3. Protozoans usually reproduce by dividing in half.

4. The metabolic rate of meiofauna averages five times that of larger animals.

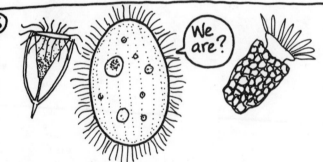

5. Ciliates are the largest group of protozoans.

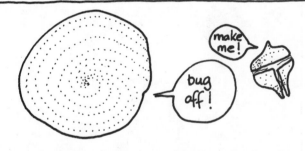

6. The largest known protozoan is less than one–half inch in diameter.

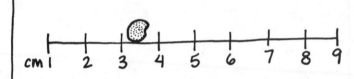

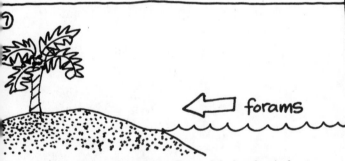

7. The pink color of the sand on Bermuda's beaches is due to crushed shells of red forams.

8. Because of their minute size, forams and radiolarians never have symbiotic organisms living within them.

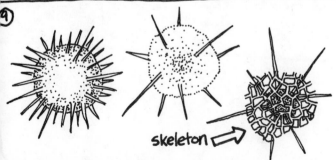

9. All radiolarians have skeletons made of silica.

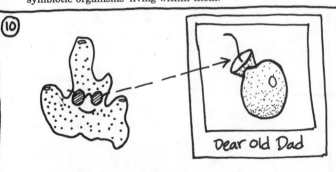

10. Protozoans may be the ancestors of sponges.

1. *True* Many biologists include dinoflagellates in the phylum Mastigophora along with other flagellated protozoans. Dinoflagellates are motile, like animals, but many of them photosynthesize, like plants. They are only one of the many organisms that cannot be definitely classified as a plant or an animal.

2. *False* Protozoans occur wherever moisture is present, including the air.

3. *True* Division is a form of asexual reproduction.

4. *True* Minute animals, like protozoa, do have high metabolic rates. Consequently they consume more food and oxygen relative to their body weights.

5. *True* There are approximately 8000 species of ciliates.

6. *False* A species of foram found in Australia may be well over half an inch in diameter.

7. *True* *Homotrema* is a large, wart-sized red foram that lives attached to coral. Its red limestone shells are a major component of Bermuda's sand.

8. *False* Both forams and radiolarians may contain symbiotic dinoflagellates (called zooxanthellae.)

9. *False* One radiolarian, *Acantharia,* has a skeleton of strontium sulfate. Strontium is an element found in minute quantities in the ocean. *Acantharia* is able to extract and concentrate strontium to form its skeleton. Many other marine organisms also have this ability to extract and concentrate a rare ingredient from their environment.

10. *True* Protozoans in the group Choanoflagellida are very similar to the collar cells (choanocytes) of sponges.

5 Sponges

Sponges (phylum Porifera) are simple animals that live permanently attached to the ocean bottom and other hard substrates. In relation to other animal groups, sponges are considered primitive. They lack distinct tissues and organs. The basic functions of animal life, such as eating, breathing, and waste removal, are carried out by individual cells acting almost independently of one another, rather than by tissues or organs. Because they have no nervous system or power of locomotion, sponges do not draw away when touched.

Their bodies are organized around a system of water canals. Two types of pores on the surface of the sponges lead to these passageways. There are many tiny incurrent pores (ostia) and a few large excurrent pores (oscula). Collar cells (choanocytes), each with a delicate collar around the base of a whip-like flagellum, line passageways throughout the sponge. The flagella beat to create water currents within the sponge, which bring in food and oxygen and remove wastes.

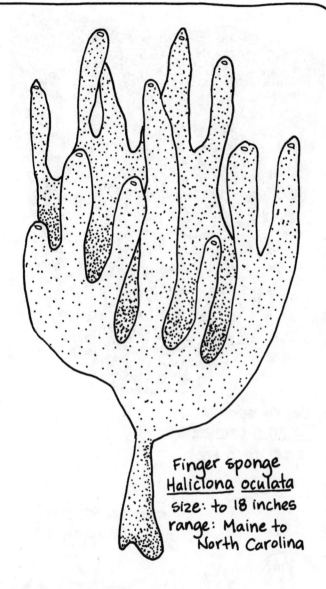

Finger sponge
Haliclona oculata
size: to 18 inches
range: Maine to
        North Carolina

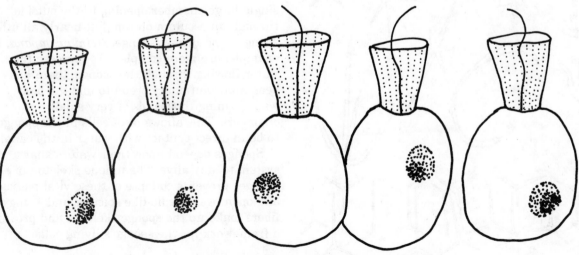

collar cells (choanocytes), greatly magnified

Sponges

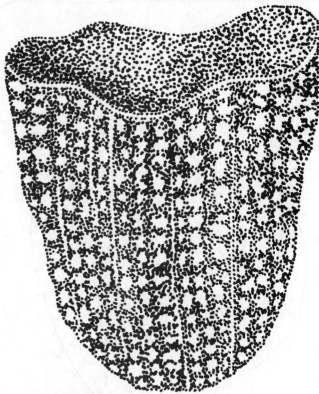

Basket sponge
Ircinia campana
size: to 3 feet

Being permanently attached to the bottom, sponges cannot pursue prey. Instead, they filter food from the water that enters through their oscula. Although it is not known exactly what sponges consume, it seems they feed on fine organic particles and small planktonic organisms in the water. Collar cells trap food particles in the water current as it passes through the canals. Studies show that 80 percent of the food sponges utilize is so minute that it cannot be seen under an ordinary microscope.

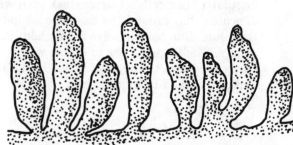

Organ pipe sponge *Leucosolenia*
size: to 1 inch

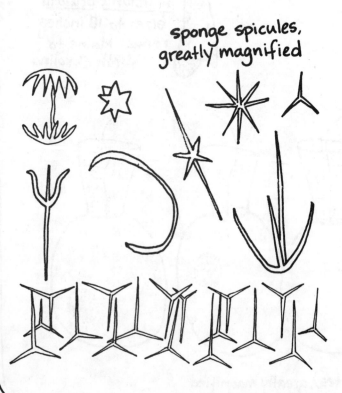

sponge spicules,
greatly magnified

Sponges vary in complexity. Simple sponges such as *Leucosolenia* have an internal cavity, a single large chamber opening to the outside through an osculum on top. But nearly all other sponges, the Basket sponge, *Ircinia campana*, for example, are more complex. To increase the water flowing through the sponge, there has been an evolutionary trend to fold the body wall, forming thousands of very small chambers. This allows most cells in the sponge to be in direct contact with water in the canals.

Sponges depend upon their water canal system to stay alive. The unique skeleton of sponges prevents collapse of these vital passages and openings. Needle-like spicules and spongy fibers make up the sponge skeleton and provide a framework for the sponge's living cells.

Sponges

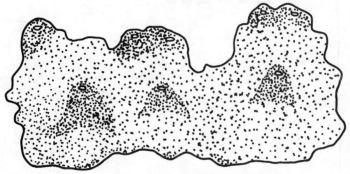

Bread crumb sponge
Halichondria panicea
size: to 1 foot across

Classifying sponges is a challenge. Most sponges are formless masses. It is not always clear whether a sponge is one animal or a colony of individuals with poorly defined body limits. To further complicate matters, the shape of a sponge is highly variable. This is true, for instance, of the Bread crumb sponge, *Halichondria panicea,* which is found in the Gulf of Maine. In the surf zone, the sponge is a thin mat, while in protected caves and tidepools, the sponge grows into thickened masses.

Positive sponge identification and classification is based on microscopic examination of skeletal structures. The skeleton may be composed of calcareous (lime) spicules, siliceous (silica) spicules, protein spongin fibers, or a combination of the latter two. The shape and composition of these skeletal structures determine the species of a sponge.

Sponges may reproduce by either sexual or asexual means. Asexual reproduction may take place by budding, in which small buds appear on the sides of body and grow into young sponges. Sponges also have remarkable powers of regeneration. Cutting a sponge in half creates two new sponges, rather then one dead one, providing the pieces become attached to the bottom. As for sexual reproduction, most sponges produce both sperm and egg. Sperm is released into the water and enters other sponges via the water currents. After fertilization, larvae develop within the parent and settle to the bottom soon after being released.

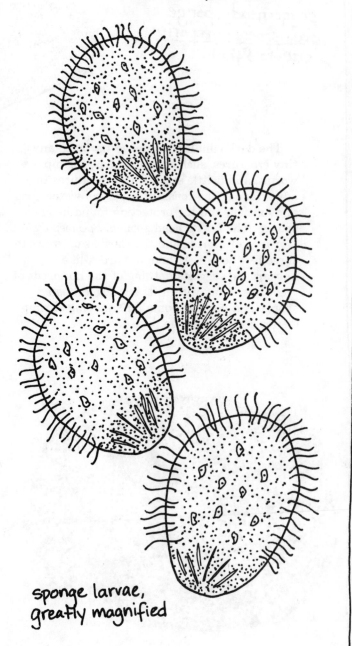

sponge larvae,
greatly magnified

51

Sponges

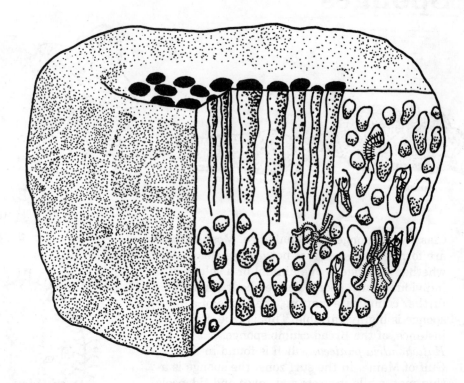

Loggerhead sponge
Speciospongia vesparia
size: to 3 feet

The dark canals of a sponge harbor many tiny creatures, such as amphipods, shrimp, worms, and brittle stars. These creatures take advantage of the constant supply of water, food and oxygen, and the protection found inside a sponge. The Loggerhead sponge, *Speciospongia vesparia,* of Florida and the Caribbean grows to three feet across. Although its rock-like appearance suggests nothing of the hundreds of creatures living within it, the Loggerhead sponge is a popular home for snapping shrimp (genus *Synalpheus,* see page 135).

The bright pink, green yellow, and blue rectangular sponges sold in stores are synthetic; they have never been alive. But real sponges are available commercially. Since the time of ancient Greece, people have recognized the absorbent, compressible, and durable nature of natural sponges. Natural sponges are suitable for many tasks, from washing dishes to packing airplane parts.

In the United States, commercial sponging takes place in Florida. The skeleton of commercial sponges is made of spongin fibers only; other types of sponges are unsuitable because of their spicules. Within the sponge, the spongin is arranged in a fibrous network that gives commercial sponges, such as the Sheepswool sponge, *Hippiospongia lachne,* the properties of resiliency and the ability to absorb large amounts of water. Through repeated squeezing, soaking, rinsing, beating, and scraping, the living parts of the Sheepswool sponge are removed, leaving behind its usable, absorbent spongin skeleton.

Sheepswool sponge *Hippospongia lachne*

Sponge True/False Quiz

1. Most sponges are a very drab color.

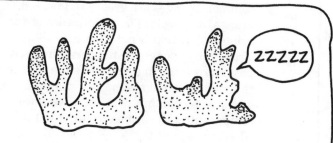

2. Sponges in northeast waters may become dormant in the winter.

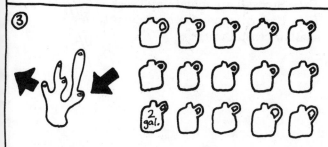

3. A sponge four inches tall and one-half inch in diameter can filter up to thirty gallons of water daily.

4. Sponges are an evolutionary dead end. It appears no animals evolved from them.

5. Sponges first appeared on earth at the time of the dinosaurs.

6. The life span of most sponges is between one and twenty years.

7. Sponges are delectable and are frequently eaten by other animals.

8. Most sponges live in fresh water.

9. Sponges are not found in the Antarctic.

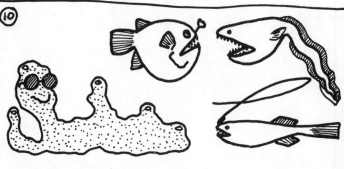

10. Most sponges live in very deep water.

1. *False* Drabness is the exception. Many common sponges are green, yellow, red, orange, or purple. The significance of the bright coloration is not known.
2. *True* Many sponges in temperate regions pass the winter in a reduced state. Parts of their water canal system deteriorate as the water gets cold, but with increasing water temperature, the sponge returns to full filtering capacity.
3. *True*
4. *True* It appears that sponges diverged early from the main line of higher animal evolution. No other members of the Animal Kingdom arose from sponges.
5. *False* Sponges are ancient animals and certainly arose before the Paleozoic Era, 550 million years ago. Dinosaurs dominated life on earth during the Mesozoic Era 180 million years ago.
6. *True* But some species may live hundreds of years.
7. *False* Sponges have few predators. Depending on the area, sponges may be eaten by seastars, nubibranchs, fish, sea urchins, or turtles, but not to a great extent.
8. *False* Most sponges are marine.
9. *False* Sponges are found in all seas.
10. *False* Sponges occur *most* abundantly in shallow coastal water.

6 Coelenterates

Almost everyone is familiar with animals in the phylum Coelenterata: hydroids, jellyfish, sea anemones, and coral. The phylum Coelenterata is divided into three classes: Hydrozoa, the hydroids; Scyphozoa, the jellyfish; and Anthozoa, the sea anemones and corals. Despite their different body shapes and lifestyles, all coelenterates have many important characteristics in common.

Coelenterates are radially symmetrical, that is, their parts are arranged around the center like spokes on a wheel. Being relatively simple animals, they have no organs, only tissues. The coelenterates are composed of two layers of cells: The outer layer is the epidermis, the inner layer is the gastrodermis. Between them is the mesoglea, a jelly-like substance. A simple digestive cavity, the coelenteron, acts as a gullet, stomach, and intestine. There is only one opening into the coelenteron, and it acts as both a mouth and an anus.

Most coelenterates have an alternation of generations, that is, they pass through two different body forms during their life cycle. One form, the polyp, has a body form like that of a sea anemone. A polyp is a cylinder with the closed end attached to the substrate, and the open end, with the mouth the tentacles, directed upward. The other body form, the medusa, is a free-swimming umbrella with the mouth opening underneath. The jellyfish is an example.

Coelenterates also contain cnidocytes, which are specialized cells used for feeding and defense. Although they are concentrated on the tentacles, cnidocytes are found in the epidermis on almost all parts of the body. Many cnidocytes have the ability to sting. This stinging ability is a unique characteristic of coelenterates. For this reason, the phylum Coelenterata is also known as the phylum Cnidaria.

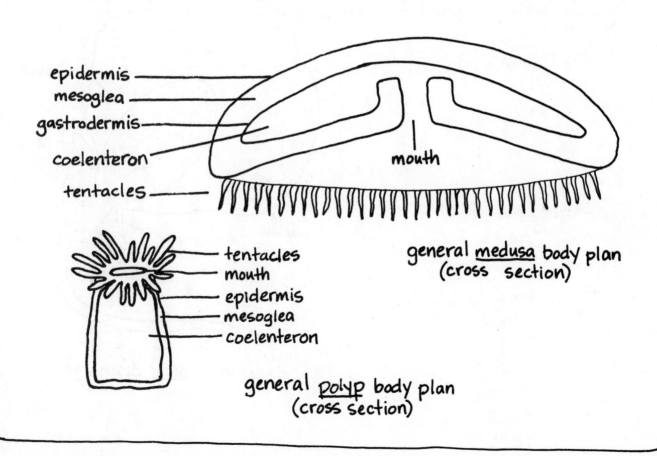

epidermis
mesoglea
gastrodermis
coelenteron
tentacles

mouth

general <u>medusa</u> body plan
(cross section)

tentacles
mouth
epidermis
mesoglea
coelenteron

general <u>polyp</u> body plan
(cross section)

Stinging Cells

Within the cnidocyte is a capsule known as a nematocyst. This capsule contains the actual stinging structure, a hollow thread with barbs lining its inside surface. How is the nematocyst released? First, the trigger at one end of the stinging cell is activated by tactile or chemical stimuli. Nerve impulses can activate the trigger as well. The pressure within the capsule forces open the lid on the cnidocyte, and the entire thread explodes outward. As the thread is discharged it turns inside out, exposing its barbs, which spring forward, flip back, and then bore into the tissue of the prey. Some nematocysts just wrap around the prey, but most inject a toxin. Nematocysts are very small, but they are effective because they are discharged in large numbers. Once the prey is captured, the coelenterate uses its tentacles to bring the prey to its mouth.
Usually the sting of a coelenterate is only strong enough to paralyze or kill tiny creatures such as zooplankton, shrimp, and small fish. Some coelenterates, such as the Portuguese man-o-war, however, are not quite as innocuous.

The nematocysts of different coelenterates vary in shape and size, and contain different toxins. They are used only once. If not pulled away by prey, they are released by the cell. New stinging cells are then formed from nearby cells.

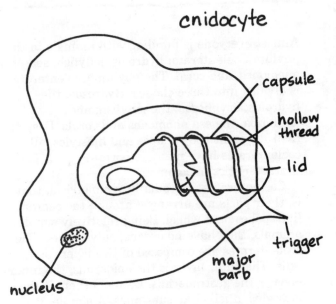

nematocyst <u>before</u> discharge

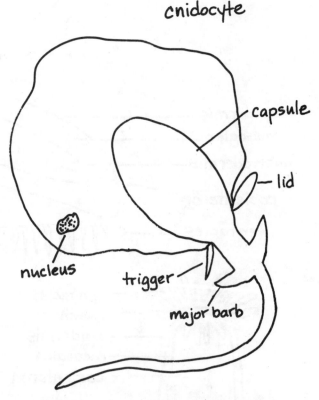

nematocyst <u>after</u> discharge

Alternation of Generations

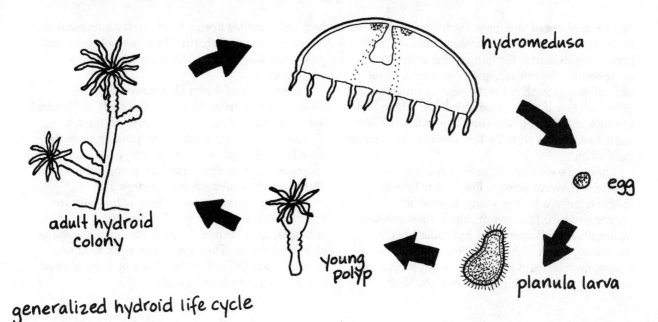

adult hydroid colony

hydromedusa

egg

planula larva

young polyp

generalized hydroid life cycle

Some coelenterates live only as polyps; others live only as medusae. Most hydroids and jellyfish (classes Hydrozoa and Scyphozoa) have an alternation of generations. At some time in their life, they pass through both polypoid (asexual) and medusoid (sexual) stages. The polyp stage is more conspicuous in hydroids; with jellyfish, the medusa is more conspicuous.

In a colony of hydroids, some of the adult polyps produce free-swimming medusae, known as hydromedusae. These jellyfish-like animals are either male or female and reproduce

sexually through the union of egg and sperm. After fertilization, the egg hatches into a microscopic larvae, the planula, which eventually settles down and metamorphoses into a polyp.

Jellyfish follow a similar developmental pattern. The adult medusa produces eggs which develop into polyps. Young jellyfish bud off the polyps. Many other hydroids and jellyfish go through variations of these basic plans. Sea anemones and corals (class Anthozoa) do not have an alternation of generations.

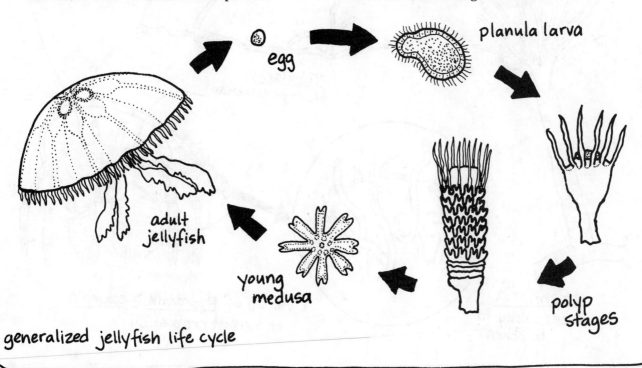

egg

planula larva

adult jellyfish

young medusa

polyp stages

generalized jellyfish life cycle

Class Hydrozoa

Most members of the class Hydrozoa, known as hydroids, are small and inconspicuous. Usually they form delicate, lacy, bush-like colonies attached to the bottom and comprise a large part of the growth on rocks, shells, and wharf pilings. They are often mistaken for algae. Despite their small size (most colonies are less than two inches wide), a few species can irritate bare skin.

There are several different types of polyps within a hydroid colony. Feeding polyps gather food. Defensive polyps sting. Reproductive polyps produce free-swimming hydromedusae, although not all species of hydroids have a medusoid stage. The polyps are all interconnected and it is hard to tell where one individual ends and another begins. Most hydroid colonies are at least partly surrounded by an envelope of chitin. Both the polypoid and medusoid stages of hydroids feed primarily on zooplankton.

Three types of hydroids found along the Atlantic coast from Maine to the Gulf of Mexico are *Tubularia*, *Pennaria*, and *Hydractinia*. *Tubularia* colonies form dense pinkish growths on pilings, jetties, and buoys. Polyps are usually unbranched, but *Pennaria* has branched polyps. *Tubularia* has no medusoid generation.

Pennaria, common on Eelgrass, has minute hydromedusae. Colonies of *Hydractinia echinata*, also known as Snail fur, encrust shells of hermit crabs. They grow as a pink fuzz and have a prediliction for the shells of Long-clawed and Flat-clawed hermit crabs.

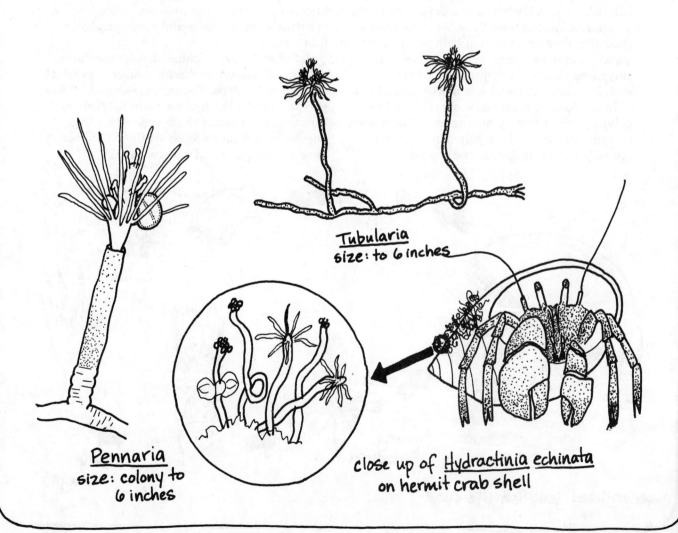

Tubularia
size: to 6 inches

Pennaria
size: colony to
6 inches

close up of Hydractinia echinata
on hermit crab shell

Siphonophores

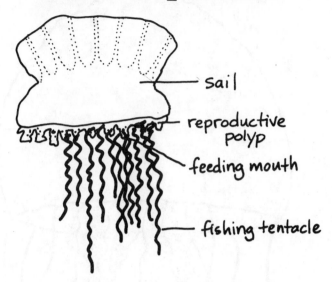

sail

reproductive polyp

feeding mouth

fishing tentacle

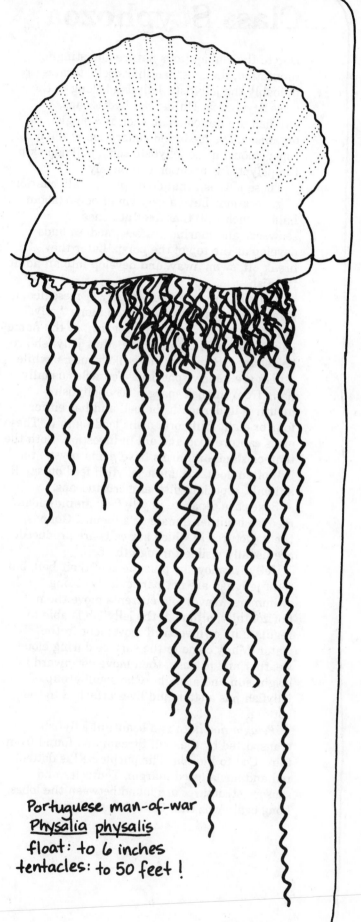

Two of the most unusual hydrozoans are the Portuguese man-of-war, *Physalia physalia,* and the By-the-wind sailor, *Velella velella.* Both are members of the order Siphonophora, in the class Hydrozoa. Like most hydrozoans, these two species are not just one animal; they are colonies of modified medusoid and polypoid individuals. Siphonophores are considered planktonic, and most species are tropical or subtropical. After storms, siphonophores can be found washed up on shore as far north as Maine.

Many siphonophores, such as the Portuguese man-of-war, have a gas-filled sac (a highly modified medusa) that acts as a float. The float of the man-of-war is a brilliant purple blue color and can reach one foot in length. Its structure enables the man-of-war to sail at a 45° angle to the wind. Tentacles, which can trail up to fifty feet behind the animal, hang from feeding polyps under the float. Powerful nematocysts on the tentacles can give bathers a severe sting. As the man-of-war has no control over its direction of travel (it moves at the mercy of the wind), it cannot be frightened away. Beached, apparently dead, specimens can still sting and should be handled cautiously, if at all.

Portuguese man-of-war
Physalia physalis
float: to 6 inches
tentacles: to 50 feet !

Class Scyphozoa

Due to their ubiquitous nature and stinging reputation, the most familiar coelenterates are probably the jellyfish. Jellyfish belong to the class Scyphozoa. Unlike the hydrozoans, jellyfish are not colonial; each is a single organism. The medusoid generation is the most conspicuous in the jellyfish; some jellyfish skip the polypoid generation completely.

Those jellyfish that do display an alternation of generations have a very small polyp (about half an inch tall) that lives attached to seaweeds and marine grasses. Medusa buds develop at the top of the polyp. Later they break off, swim away, and develop into an adult jellyfish.

Jellyfish are composed largely of mesoglea, a gelatinous substance, hence the name "jelly." Scyphozoans, however, are not fish, so the name jelly*fish* seems inappropriate. Most jellyfish are bell-shaped. Some are more like saucers, while others are nearly spherical. Tentacles usually hang from the bell margin. The jellyfish's mouth is underneath the bell and has either four or eight frilly oral arms leading to it. These arms capture food and aid in ingestion. Both the tentacles and oral arms have nematocysts; the bell often does, too. Most jellyfish feed on small animals, especially fish and crustaceans.

Sexes are separate in jellyfish; an individual medusa is either a male or a female. Gonads, the structures in which sex cells are produced, are usually visible through the bell.

Jellyfish move by pulsations of their bell, but most jellyfish are not particularly strong swimmers. Waves and currents move them horizontally, although the jellyfish is able to regulate its own vertical movement in the water. Many come to the surface during cloudy weather and at dusk, then move downward in bright sun and at night. One small group of jellyfish has a stalk and lives attached to the bottom.

Pelagia noctiluca is a beautiful jellyfish transported by the Gulf Stream and found from Cape Cod to Florida. The purple bell is dotted red and has a lobed margin. Tentacles and sensory structures are found between the lobes. Long oral arms trail behind the jellyfish.

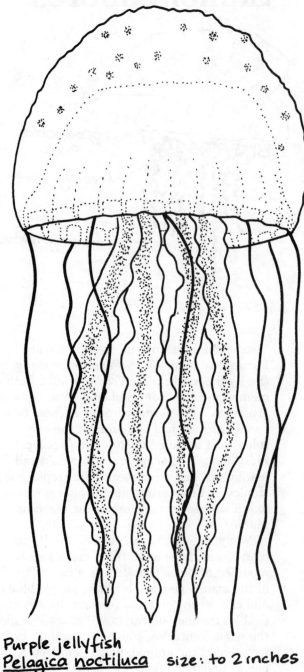

Purple jellyfish
Pelagica noctiluca   size: to 2 inches

Moon Jelly and Cannonball Jellyfish

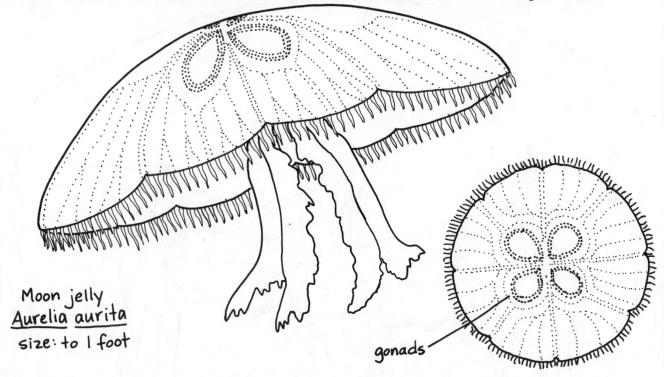

Moon jelly
Aurelia aurita
size: to 1 foot

gonads

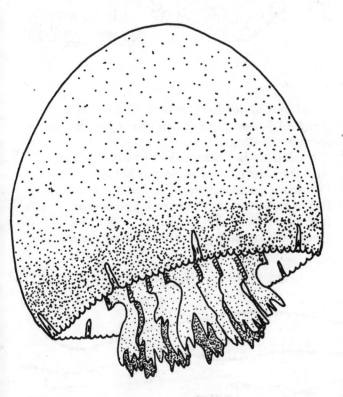

Cannonball jellyfish
Stomolophus meleagris
size: to 1 foot

The moon jelly, *Aurelia aurita,* is one of the most common jellyfish along the Atlantic Coast and is almost worldwide in distribution. Usually it is a translucent white, pink or beige color. Its four horseshoe-shaped gonads are conspicuous. About two hundred and fifty hair-like tentacles hang from its bell margin. The sticky mucus-covered tentacles catch small planktonic organisms, and the oral arms wipe off this food and bring it to the mouth.

Although the moon jelly is slightly venomous, contact with it produces at most a temporary burning sensation. The clear jelly-like blobs that wash up on beaches are often dead moon jellies.

The Cannonball jellyfish, *Stomolophus meleagris,* has a thick muscular bell resembling half an egg. There are no tentacles around its bell, but its eight oral arms join together to form a tube-like mouth. Although rarely found north of Chesapeake Bay, the Cannonball jellyfish is very common from North Carolina to Florida.

Lion's Mane and Sea Nettle

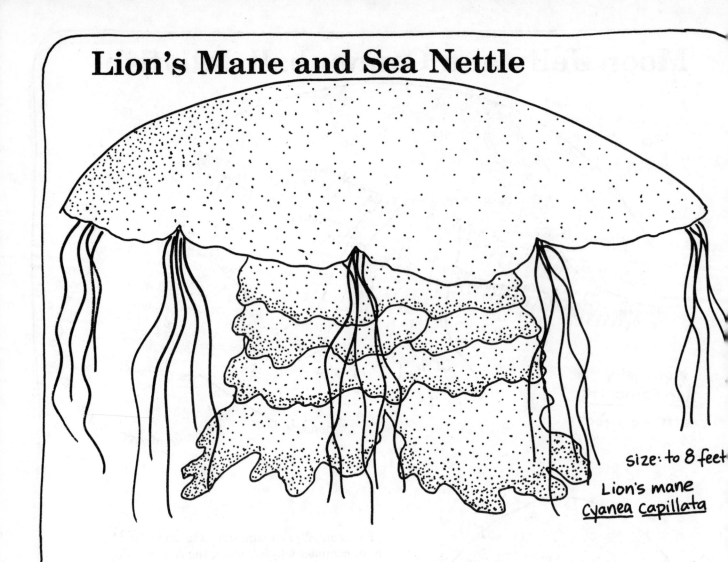

size: to 8 feet

Lion's mane
Cyanea capillata

With a bell that can grow to eight feet in diameter (although wider than three feet is rare), the Lion's mane jellyfish, _Cyanea capillata_, is the world's largest jellyfish. It can grow from a single one-eighth-of-an-inch immature medusa to an eight-foot-wide adult in _one_ summer. Lion's mane jellyfish are found from Maine to Florida. Their color and size vary with the habitat. The biggest specimens are found in cold northern waters, where food is plentiful. These jellyfish can give a painful, but short-lived, sting.

Swimmers in Cheaspeake Bay are familiar with the sharp stings of the Sea nettle, _Chrysaora quinquecirrha_. The Sea nettle can live in salt water as dilute as 3 ppt., and is found from Cape Cod to Florida. Marine specimens are larger than those in estuaries.

Sea nettle _Chrysaora quinquecirrha_
size: to 7.5 inches

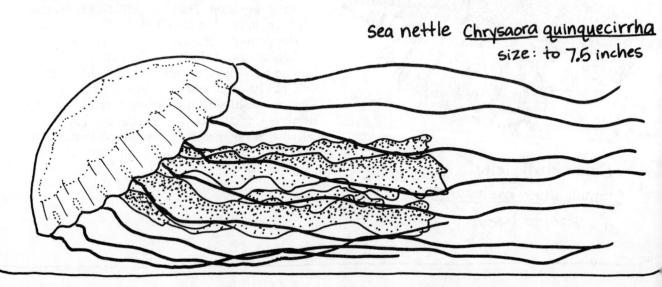

Class Anthozoa

Sea anemones, sea fans, and corals are coelenterates belonging to the class Anthozoa. Anthozoans have no alternation of generations. They live either as solitary polyps, such as sea anemones, or in colonies of polyps, as do the corals.

Like all polyps, anthozoans have a relatively simple body plan. Their body is a hollow tube open at one end, the mouth. Tentacles covered with cnidocytes (stinging cells) surround the mouth. The coelenteron (gut) of anthozoans is partitioned into radiating compartments by structures known as septa. These partitions increase the amount of surface area in the gut, allowing more efficient digestion.

Sea anemones are solitary polyps. Most live attached to rocks, shells, or submerged objects, but a few burrow in sand or mud. Although they are essentially sessile, sea anemones can change locations, albeit slowly. They may glide along on their basal disc or crawl on their side. Some even make valiant attempts at swimming by lashing their tentacles. Burrowing anemones dig themselves into mud with muscular contractions of their body.

Tentacles of sea anemones can be extended, in which case the anemone will be firm and erect, or contracted, in which case the anemone will resemble a stewed tomato. The anemone uses its tentacles to sting and trap food, generally plankton and various invertebrates. The tentacles also serve a defensive function, but their sting is usually not strong enough to be felt by people.

Anemones can reproduce sexually, releasing eggs and sperm, but they also have many innovative methods of asexual reproduction. As they creep around, they may leave behind fragments of tissue which may grow into adult anemones. Many pull apart into two halves, as illustrated on this page.

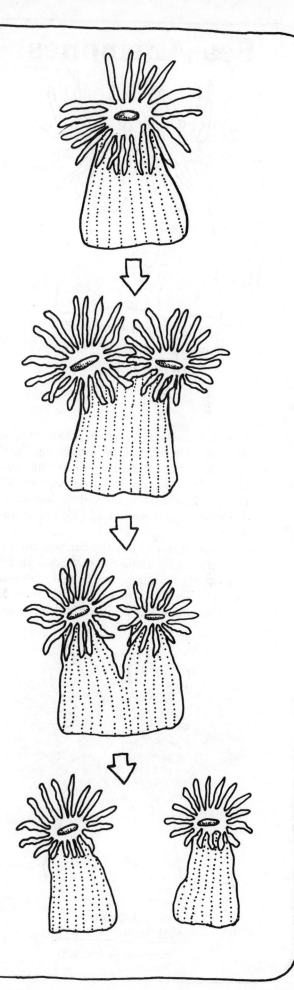

Sea Anemones

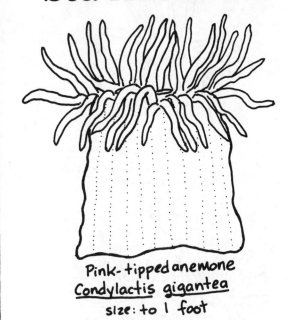

Pink-tipped anemone
Condylactis gigantea
size: to 1 foot

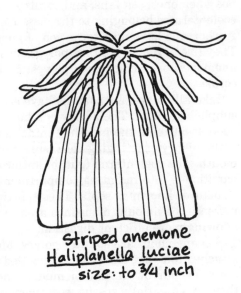

Striped anemone
Haliplanella luciae
size: to ¾ inch

Sea anemones are found in a variety of marine habitats. The Pink-tipped anemone, _Condylactis gigantea_, attaches itself to rocks and hard objects in reefs and shallow grass beds along the coast of Florida. Commensal shrimp such as _Periclimenes pedersoni_ (pg. 135) are often found with it.

Haliplanella luciae, the Striped anemone, was first found along the east coast in 1892, apparently an immigrant from Japan and Europe. Now it grows abundantly from Maine to Chesapeake Bay on intertidal rocks and pilings. It has a green column with red or yellowish stripes.

The largest and most common anemone north of Delaware Bay is _Metridium senile_, the Frilled anemone. Compared to other species of sea anemones, its tentacles are fine and more numerous. It grows intertidally as well as below the low tide mark.

Actinothoe modesta is a long worm-like burrowing anemone found from Cape Cod to Cape Hatteras. When disturbed, it withdraws completely into the sand.

Frilled anemone
Metridium senile
size: to 4 inches

Burrowing anemone
Actinothoe modesta
size: to 2½ inches

Marine Symbiosis

Over the course of evolution, many marine plants and animals have improved their chances of survival by developing a close association with some other organism. Most marine organisms spend at least part, if not all, of their lives living on or with another organism. These types of relationships are called *symbiotic.*

The word symbiosis, derived from the Greek term meaning "to live together," has several meanings. Here, symbiosis will be defined as two dissimilar organisms living in close association, *regardless* of the harm or benefit to either. Symbiotic associations can be between two plants, two animals or a plant and an animal.

In some symbiotic relationships, the two organisms are physiologically dependent on one another. In other symbiotic relationships, the organisms could just as easily live apart. Although it is difficult to categorize symbiotic associations, three general types are recognized:

1. *mutualism:* both organisms benefit from living together
2. *commensalism:* one organism benefits, one is unaffected
3. *parasitism:* one organism benefits, one is harmed

The reasons for symbiosis are also hard to pinpoint. An association may involve one or more of these elements:

1. protection
2. food
3. cleaning
4. transportation

Sea anemones are partners in many symbiotic asociations. One of the best known examples occurs in the Red Sea and Indo-Pacific between the sea anemone *Premnas* and the clownfish *Amphiprion.* Fishes the size of the clownfish are usually stung and eaten by the anemone. However, through a gradual acclimation process, the clownfish becomes immune to the anemone's nematocysts. The fish repeatedly brushes against the tentacles until its own mucus coating produces a special chemical that inhibits the anemone's stinging. The anemone provides protection and food scraps for the fish. In turn, the fish lures other creatures into the anemone's tentacles and may also remove dead and dying tissue from the anemone. This relationship is mutualistic, since both the anemone and the fish benefit from it.

Another symbiotic association occurs between certain anemones (especially *Calliactis)* and hermit crabs (especially *Pagurus).* In this relationship the anemone feeds on scraps of food that float away as the hermit crab eats. In turn, the anemone provides protection for the hermit crab.

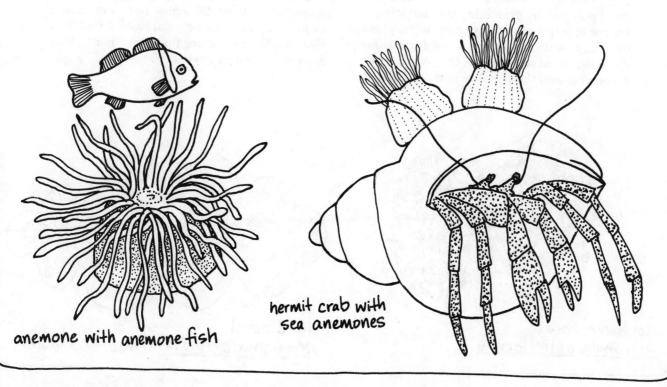

anemone with anemone fish

hermit crab with sea anemones

Corals

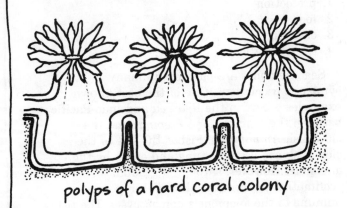

polyps of a hard coral colony

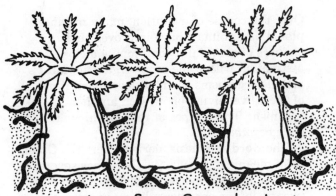

polyps of a soft coral colony

Corals belong to the class Anthozoa. Like sea anemones, coral polyps have stinging cells (usually their sting is only strong enough to paralyze the plankton on which they feed) and *no* alternation of generations. But corals differ from anemones in two important ways: they are generally colonial, and they secrete a skeleton. In other words, corals are colonies of polyps that live together within a common skeleton. A thin skin-like layer over the skeleton connects each polyp to its neighbors. Hard, or stony, corals secrete a calcareous limestone skeleton. Reef-building corals are hard corals. Soft corals, such as sea fans and sea whips, secrete a soft flexible skeleton. A sure way to tell the difference between soft coral polyps and hard coral polyps is by examining the tentacles. Polyps of soft corals always have eight pinnate (feathery) tentacles; polyps of hard corals never do. When coral colonies die, the polyps decompose and the skeleton remains behind.

Not all corals build reefs. Only areas with warm clear water can support coral reef formation. But coral is found all along the Atlantic coast from Maine to Florida.

Astrangia astreiformis, known as Star or Northern coral, is the only hard coral likely to be seen while snorkeling off the northeast coast. As in all hard corals, the polyp secretes a skeletal cup within which it lives. Actually, the coral colony lies completely above the skeleton it secretes. To protect themselves, polyps can withdraw almost completely into the cup. When the polyps extend their tentacles, they form a mesh, which snares passing plankton.

When the coral polyps are well separated from each other the colony has a pitted appearance, as in *Astrangia.* But in corals such as brain coral and rose coral (both found in Florida), the polyps are fused together in long rows separated by skeletal ridges.

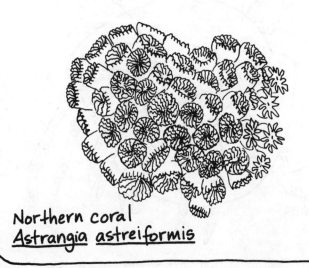

Northern coral
Astrangia astreiformis

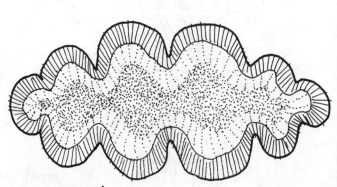

Rose coral
Manicina areolata   size: to 6 inches

Reef-building Corals

Coral reefs are produced by millions of coral polyps, each removing calcium and carbonate from the sea water and depositing it as a hard skeleton. As one generation of polyps reproduces, the new polyps build their skeleton on the foundation of the old. This cycle has resulted in the build-up of coral into huge limestone reefs.

Coral reefs are found between 25° north latitude and 25° south latitude where the water is clear, sunlit, and warm. Reefs are important land builders, as evidenced by the Florida Keys, which are actually ancient coral reefs left high and dry through changing sea levels during the Ice Ages.

All reef-building corals possess symbiotic algae within their tissues. These abundant microscopic algae (up to 30,000 algae per cubic millimeter of coral tissue) are called zooxanthellae. Zooxanthellae are actually dinoflagellates (see page 18) in an encysted state. Their yellow-brown pigments give the coral its color. It is largely the symbiotic relationship between zooxanthellae and coral that makes reef formation possible.

Within the cells of the coral polyps, zooxanthellae live, reproduce, and photosynthesize, utilizing the waste products of the coral (carbon dioxide, nitrogen, and phosphorous). In turn, the coral may utilize food and oxygen produced by the algae during photosynthesis. Without the zooxanthellae, the corals would not have the ability to secrete the massive stony skeleton we call the reef.

Since zooxanthellae need light for photosynthesis, reef-building corals grow only in ocean waters less than 300 feet deep. The corals need warm water (above 70°F) and do not tolerate low salinity or murky water. The only place in the continental United States suitable for coral reef formation is around the Florida Keys, where there is warm, clear water from the Gulf Stream.

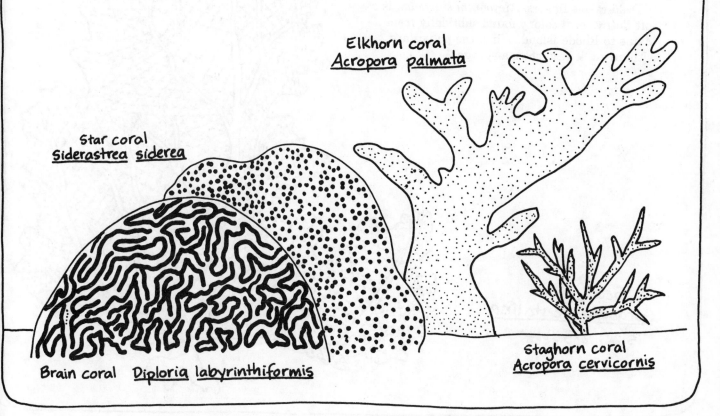

Star coral
Siderastrea siderea

Elkhorn coral
Acropora palmata

Brain coral   Diploria labyrinthiformis

Staghorn coral
Acropora cervicornis

Soft Corals

Polyps of soft corals have eight pinnate tentacles. For this reason soft corals are also called octocorals. Both hard and soft coral polyps are small, usually only a couple of millimeters across, but easily visible to the naked eye.

A fleshy mass of tissue connects all the soft coral polyps together in a colony. This tissue contains mesoglea and gastrodermal tubes, which join the digestive cavities of all the polyps together. Over the surface of this fleshy tissue is an epidermal layer, which joins the epidermis of the polyps. Cells in the tissue secrete a skeleton inside the colony. The skeleton is flexible and made of a horn-like material or calcium carbonate spicules.

Whip corals and sea fans belong to the group of soft corals called Gorgonians, named for the central axis of their skeleton, which is made of protein called gorgon. They look rather plant-like, but their branching shapes provide extensive surface area for catching plankton as it drifts by. Sea fans are found in the Florida Keys.

Sea pansies, *Renilla reniformis*, are soft corals that look like lilypads but they are actually one large primary polyp (with a stalk) covered with secondary polyps on the upper side. Sea pansies grow subtidally on sandy bottoms from North Carolina to Florida.

Dead man's fingers, *Alcyonium sidereum*, is a soft fleshy coral colony found subtidally from Maine to Rhode Island. When the polyps are withdrawn, it resembles a sponge.

Sea pansy
Renilla
reniformis
size: to 2 inches

Dead man's fingers
Alcyonium digitatum size: to 8 inches

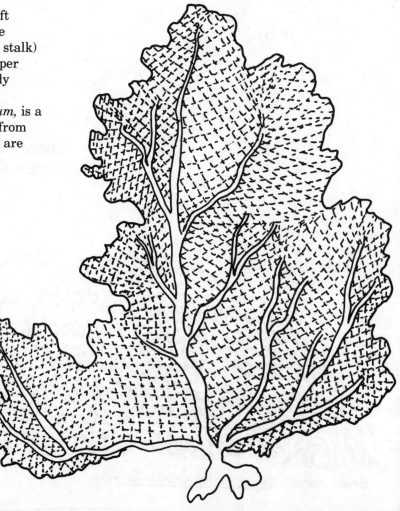

Sea fan
Gorgonia ventalina
size: 2-3 feet

Coelenterate True/False Quiz

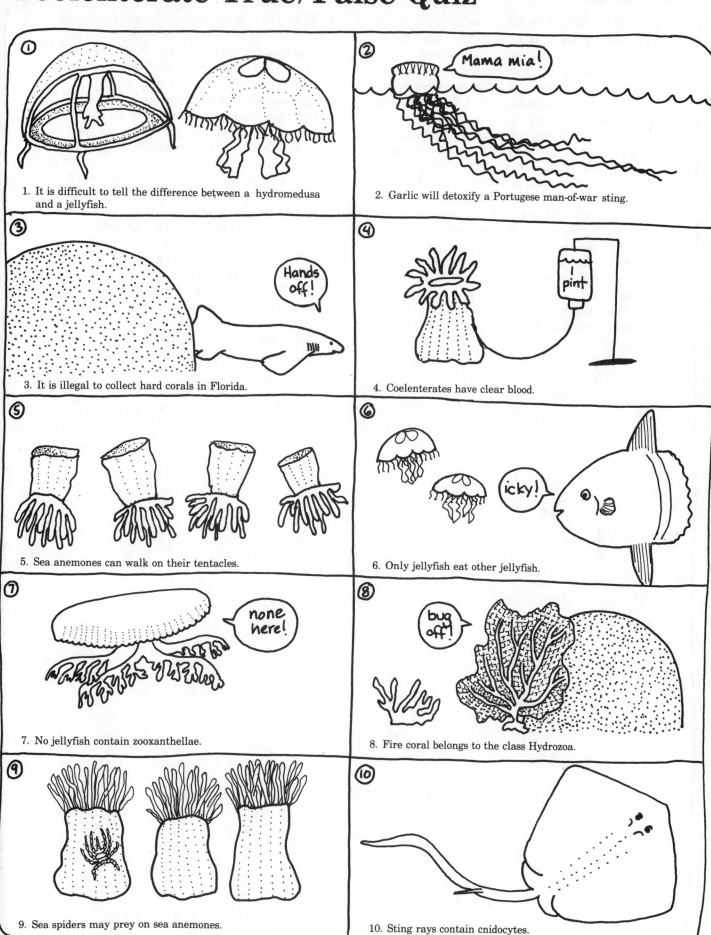

1. It is difficult to tell the difference between a hydromedusa and a jellyfish.

2. Garlic will detoxify a Portugese man-of-war sting.

3. It is illegal to collect hard corals in Florida.

4. Coelenterates have clear blood.

5. Sea anemones can walk on their tentacles.

6. Only jellyfish eat other jellyfish.

7. No jellyfish contain zooxanthellae.

8. Fire coral belongs to the class Hydrozoa.

9. Sea spiders may prey on sea anemones.

10. Sting rays contain cnidocytes.

1. *False* Unlike jellyfish (class Scyphozoa), the hydromedusae (class Hydrozoa) are usually small, from one-quarter inch to two inches in diameter. A major anatomical difference is also obvious: On the hydromedusae, the bell margin projects inward, forming a shelf known as the velum.
2. *False* Garlic won't, but meat tenderizer will! An enzyme in the meat tenderizer breaks down the protein in the nematocyst's toxin.
3. *True* In Florida it is unlawful to take, possess, or destroy sea fans, hard corals, or fire corals unless it can be shown by certified invoice that it was imported from a foreign country.
4. *False* Coelenterates have no blood.
5. *True* But sea anemones do not walk with celerity.
6. *False* Many creatures, including the Giant ocean sunfish, *Mola mola,* (pictured) eat jellyfish.
7. *False* The Upside-down jellyfish, *Cassiopeia xamachana,* a common jellyfish in Florida, does contain zooxanthellae. It lives upside down (oral arms up) in quiet shallow water soaking up sunshine, allowing the zooxanthellae to photosynthesize. The Upside-down jellyfish can catch its own food, but in adequate light it can survive primarily on the food products produced by the zooxanthellae.
8. *True* Fire coral, *Millepora,* belongs to the class Hydrozoa, not Anthozoa. It is colonial and secretes a hard skeleton, making it look like a hard coral, but it isn't. Defensive polyps emerge from pores in the skeleton and can produce a stinging sensation if crashed into at high speeds.
9. *True* Sea spiders are carnivorous and may eat the anemone on which they live.
10. *False* Only coelenterates produce cnidocytes. The sting ray is a fish, and it "stings" with a barb at the end of its tail.

7 Comb Jellies

Comb jellies (phylum Ctenophora) are fragile, beautiful animals. They are often mistaken for jellyfish. Both jellyfish and comb jellies are gelatinous and have a similar medusoid body plan. But there are major differences between the two.

The body of a comb jelly is divided into eight equal sections by eight bands of hair-like cilia. These bands, called comb rows, are the main characteristic of the phylum Ctenophora. Each band extends from the aboral end almost all the way to the oral end. The comb rows all beat together, providing locomotor powers to the comb jelly. By beating in waves from the oral end toward the aboral end, the comb rows move the comb jelly mouth-first through the water. When the comb jelly encounters an object, the beat can be temporarily reversed, allowing the comb jelly to back away.

Most comb jellies cannot sting. Only one species, not found on the east coast, has cnidocytes. Comb jellies catch their food with their tentacles and/or lobes. Some comb jellies, the Sea gooseberry for example, have two tentacles emerging from canals near the aboral end. These branched tentacles are covered with adhesive cells and form a very effective net when expanded. Plankton are trapped on the sticky tentacles and are wiped off the tentacles into the comb jelly's mouth. Sea gooseberries prey heavily upon fish eggs and larvae.

Species of *Mnemiopsis* use tentacles and their mucus-covered lobes to trap food. *Beroë* comb jellies have no tentacles, but do have a cavernous mouth with which they can swallow other comb jellies whole.

Comb jellies contain both male and female structures. Eggs are fertilized externally in the water. There is no alternation of generations; the larva develops directly into a juvenile comb jelly.

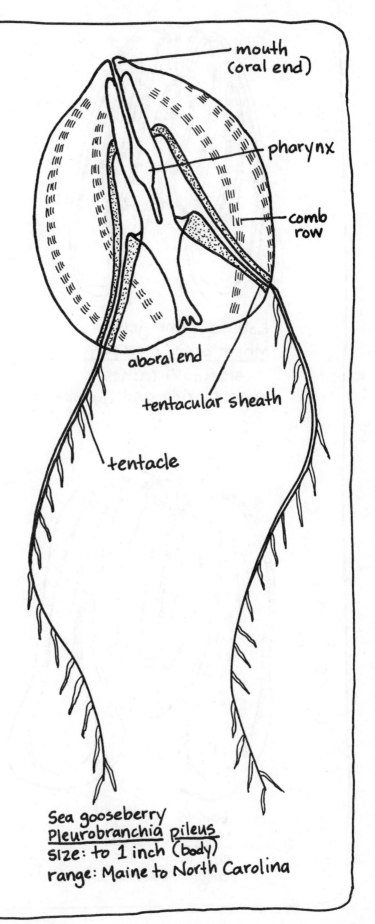

Sea gooseberry
Pleurobranchia pileus
size: to 1 inch (body)
range: Maine to North Carolina

Comb Jellies

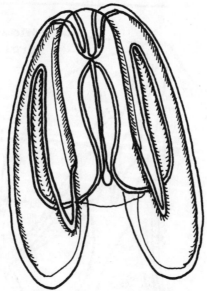

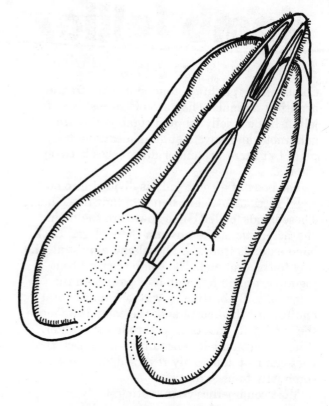

Lobate comb jelly
<u>Mnemiopsis</u> <u>mccradyi</u>
size: to 4 inches
range: Florida and Caribbean

Common northern comb jelly
<u>Bolinopsis</u> <u>infundibulum</u>
size: to 6 inches
range: north of Cape Cod

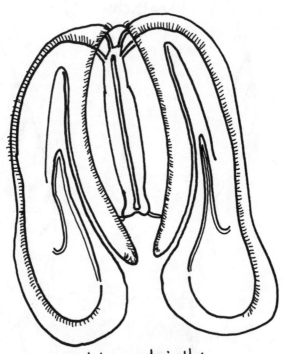

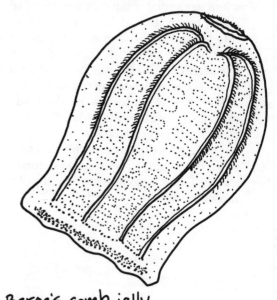

Leidy's comb jelly
<u>Mnemiopsis</u> <u>leidi</u>
size: to 4 inches
range: south of Cape Cod

Beroe's comb jelly
<u>Beroe</u>
size: to 4½ inches
range: south of Cape Cod

Comb Jelly True/False Quiz

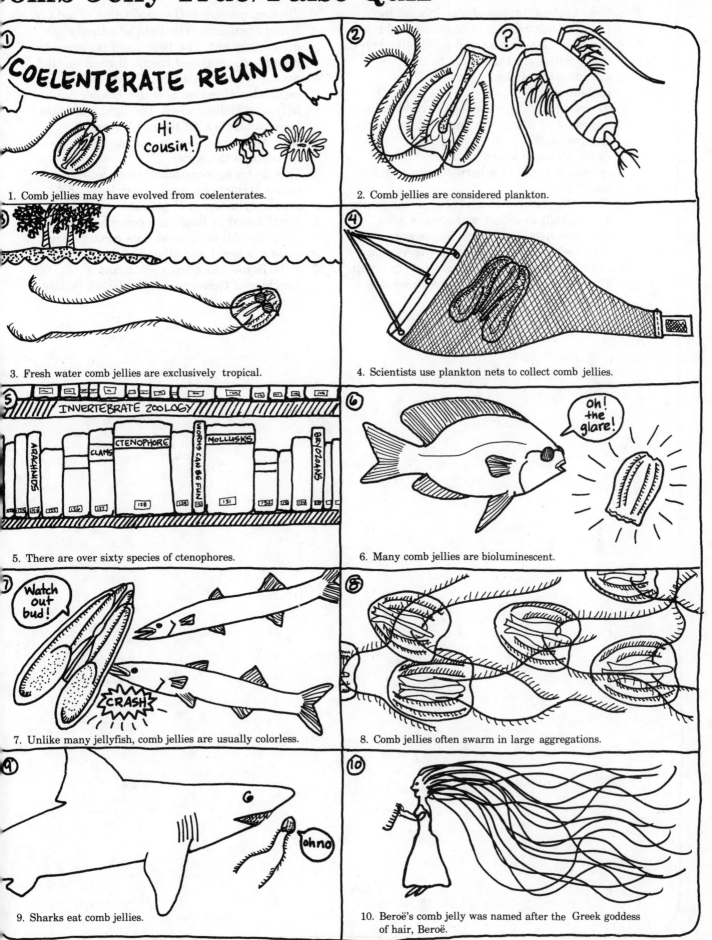

COELENTERATE REUNION

Hi cousin!

1. Comb jellies may have evolved from coelenterates.

2. Comb jellies are considered plankton.

3. Fresh water comb jellies are exclusively tropical.

4. Scientists use plankton nets to collect comb jellies.

INVERTEBRATE ZOOLOGY

ARACHNIDS CLAMS CTENOPHORE WORMS CAN BE FUN MOLLUSKS BRYOZOANS

5. There are over sixty species of ctenophores.

Oh! the glare!

6. Many comb jellies are bioluminescent.

Watch out bud! CRASH

7. Unlike many jellyfish, comb jellies are usually colorless.

8. Comb jellies often swarm in large aggregations.

oh no

9. Sharks eat comb jellies.

10. Beroë's comb jelly was named after the Greek goddess of hair, Beroë.

1. *True* Coelenterates and ctenophores have many characteristics in common, including a medusoid body plan and a jelly-like layer. Comb jellies are thought to be an evolutionary offshoot from the coelenterates.
2. *True* Although they can swim, comb jellies are more or less at the mercy of the waves and currents.
3. *False* All comb jellies are marine. No species, tropical or otherwise, live in fresh water.
4. *False* Due to their fragile body, comb jellies are difficult to collect undamaged with plankton nets. Most comb jellies are collected by divers who catch them in jars.
5. *False* The phylum Ctenophora is very small. There are only about fifty known species of comb jellies.
6. *Bioluminescence* is the production of light by living organisms. The light of a firefly is bioluminescent. The light itself is caused by an enzyme-catalyzed chemical reaction that produces very little heat. Comb jellies, especially *Mnemiopsis leidyi,* Leidy's comb jelly, can bioluminesce. The light emanates from canals behind the comb rows.
7. *True* Most are colorless and are nearly invisible in the water.
8. *True* As an aggregation of comb jellies moves through an area, it can leave the area nearly devoid of plankton.
9. *True* The spiny dogfish, a common shark along the Atlantic coast, eats comb jellies, as do mackeral and other fish.
10. *False* Beroë was a sea nymph and a daughter of Oceanus, not the Greek goddess of hair.

8 Marine Worms

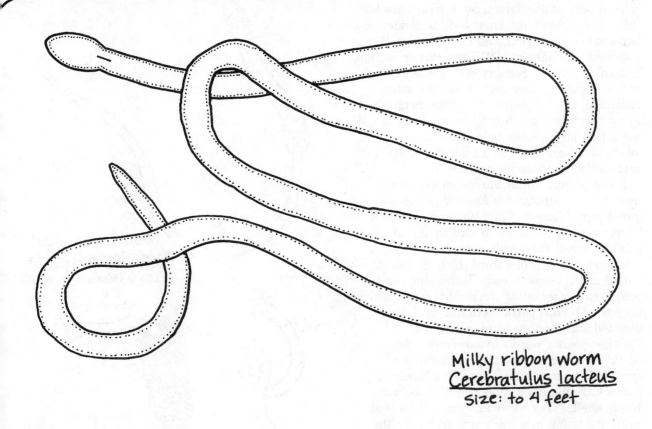

Milky ribbon worm
Cerebratulus lacteus
Size: to 4 feet

Marine worms are a very diverse group of animals. There are at least six different phyla (singular: phylum) of worms, and other than their worm-like shape, they have little in common.

If you have ever found a flat, soggy, noodle-like creature on the shore that broke into pieces when you picked it up, you know what a ribbon worm is. Most ribbon worms live on the bottom beneath shells, stones, and seaweed. Some burrow into mud and sand. Using muscular contractions and cilia, ribbon worms glide over the bottom on a trail of slime. Nighttime is when they are the most active; they avoid light.

Unlike many other marine worms, ribbon worms have no appendages and no segments. Many, however, have simple eye spots on the underside of the anterior ("head") end. The mouth, which is nothing more than a long slit, is also underneath the head.

What makes ribbon worms unique is their proboscis, a long, tongue-like organ literally shot out of the body from a pore near the mouth. Ribbon worms use their proboscis for defense, burrowing, and prey capture. It is not used to impale human beings. When discharged, the proboscis turns inside out and can protrude far beyond the body. Ribbon worms are carnivorous and feed on small worms, mollusks, and crustaceans. Food items are grabbed by the proboscis, sucked quickly into the worm's mouth, and swallowed whole.

No matter how gingerly ribbon worms are handled, they inevitably break apart. But the pieces can live independently and regenerate parts.

The Milky ribbon worm, *Cerebratulus lacteus*, which is common along the whole coast, can grow as long as four feet and up to five-eighths of an inch wide. But many ribbon worms are much smaller, growing only as long as an inch or two.

Peanut Worms

Peanut worms (phylum Sipunculida) are drab-colored bottom dwellers, most of which are less then four inches long. Their body is divided into two sections: the trunk and the introvert. The introvert, the narrow neck-like anterior section, is used as a probe. Sensory cells are abundant at the end of the introvert. When the worm is disturbed, it pulls its introvert into its trunk. Peanut worms are sturdy little creatures and do not fall to pieces when held. Their mouth, located at the end of the introvert, is surrounded by tentacles or lobes.

Some peanut worms burrow in sand and mud. Gould's sipunculid, *Phascolopsis gouldi,* is found from Maine to Cape Hatteras, where it burrows along the shore in intertidal and subtidal areas. It is easily recognized by its introvert, which is one third the length of its trunk, and its smooth body. Burrowing peanut worms eat by ingesting the sand through which they travel. Food particles in the sand are digested and then the sand is expelled.

Other peanut worms live in empty snail shells. The Hermit sipunculid, *Phascolion strombi,* lives subtidally from Nova Scotia to North Carolina. Its body is modified to fit the empty shells. Tiny hooks and projections that circle the trunk help the worm grab onto the inside of the shell. The Hermit sipunculid draws in sediment with its tentacles and digests food particles found among the sand grains.

A few peanut worms even bore into rocks or coral. With the small teeth on its introvert, the Antillean sipunculid, *Phascolosoma antillarum,* burrows in rocks along the shore of Florida and the Caribbean. In addition to using its teeth, it may also exude a chemical which helps break down the rock. By opening its tentacles at the mouth of its burrow, the Antillean sipunculid creates a current and may draw food toward itself.

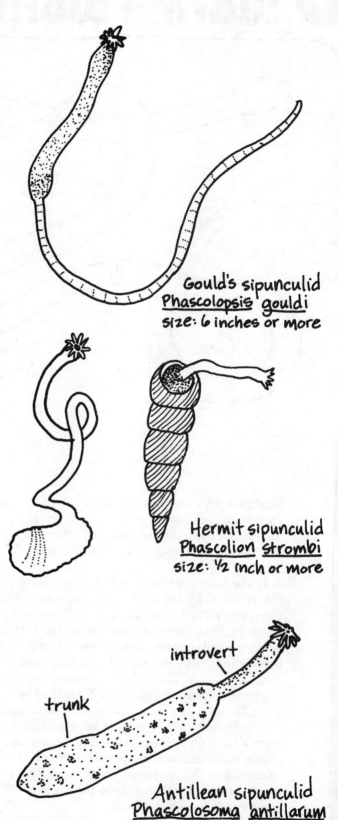

Gould's sipunculid
Phascolopsis gouldi
size: 6 inches or more

Hermit sipunculid
Phascolion strombi
size: 1/2 inch or more

introvert

trunk

Antillean sipunculid
Phascolosoma antillarum
size: 2 inches

Acorn Worms

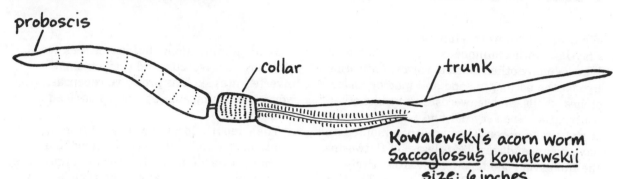

proboscis

collar

trunk

Kowalewsky's acorn worm
Saccoglossus kowalewskii
size: 6 inches

Compared to other marine worms and most other invertebrates, acorn worms (phylum Hemichordata) are highly specialized and evolutionarily advanced. Because of their gill structure, which is similar to some primitive fish-like animals, acorn worms are often considered to be one of the links between vertebrates and invertebrates.

Acorn worms are rather lethargic, shallow water creatures. Some species live under shells and stones; other burrow. All acorn worms have fragile, flaccid bodies composed of three sections: the proboscis, the collar, which contains the mouth, and the long trunk, which bears a row of gill pores. They are almost impossible to collect intact. Most acorn worms can regenerate at least part of their trunk.

Kowalewsky's acorn worm, *Saccoglossus kowalewski,* is found from Maine to North Carolina in intertidal and subtidal mud flats. It constructs a U-shaped, mucus-lined burrow with two openings to the surface. One or both ends of the worm may stick out of the openings. To burrow, the acorn worm anchors its proboscis, then pulls its body towards it through muscular contractions.

As the worm burrows, it consumes mud and sand from which it digests food particles. At the opening to the burrow, small (one-half-inch) piles of stringy mud castings accumulate, indicating the huge amount of mud passed through the worm's body. Look for piles of these castings on sand and mud flats.

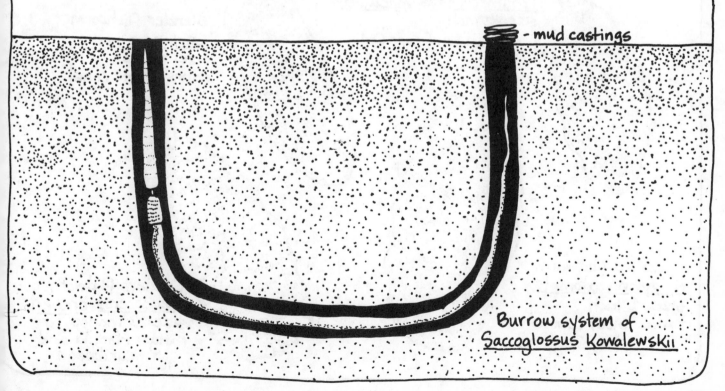

- mud castings

Burrow system of
Saccoglossus kowalewskii

Flat Worms

What makes flatworms (phylum Platyhelminthes) unique is not their habitat. Like so many other marine worms, flatworms live on the bottom in sand and mud or under stones, shells, and seaweeds. But unlike other worms, they are very flat and resemble a piece of wet tissue paper. They are unsegmented and have no appendages. Two classes of flatworms, the flukes and the tapeworms, are entirely parasitic, and are not discussed here. The third class contains the turbellarians, which are primarily free-living and very common on the shore. Many are minute and live among sand grains.

Flatworms glide over the bottom, using hair-like cilia on their underside for propulsion. Some can swim with undulating muscular contractions. Numerous adhesive gland cells in or under the epidermis help the larger worms to move smoothly over the bottom and the smaller ones to adhere to sand grains. Their flat shape affords them greater contact with the substrate.

While gliding along, flatworms seek food. Most are carnivorous, and prey on small invertebrates such as protozoa, copepods, worms, and tunicates, as well as on dead animals.

Their mouth, located well back from the front, on the underside, takes in food and discharges waste. Many flatworms capture prey by wrapping themselves around their victim, entangling it in slime and sticking it to the substrate with an adhesive substance. Then the tube-like proboscis is extended from their mouth to the prey, and digestive juices are poured onto it. The flatworm sucks the semi-liquified meal into its body. Flatworms can feed on large chunks of dead meat this way. Prey may also be swallowed whole.

Most flatworms have eyes, usually two, but four or six are not uncommon. Their eyes function only to detect light, which the flatworms avoid.

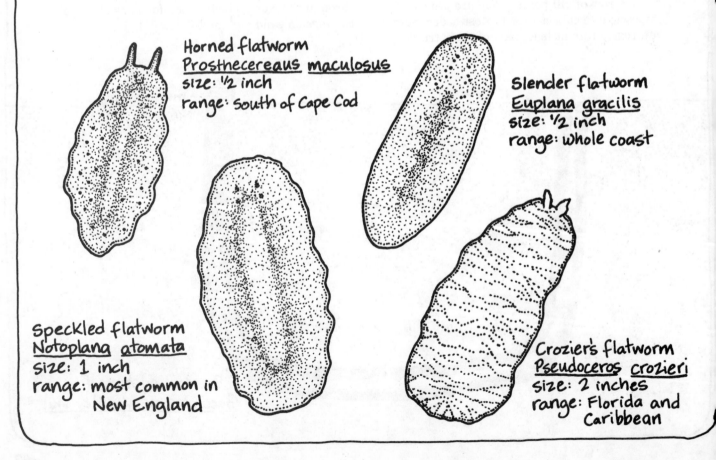

Horned flatworm
Prosthecereaus maculosus
size: ½ inch
range: south of Cape Cod

Slender flatworm
Euplana gracilis
size: ½ inch
range: whole coast

Speckled flatworm
Notoplana atomata
size: 1 inch
range: most common in New England

Crozier's flatworm
Pseudoceros crozieri
size: 2 inches
range: Florida and Caribbean

Flat Worms

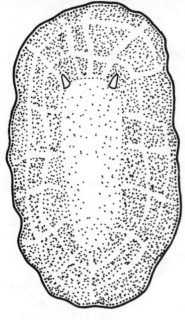

Oyster flatworm
Stylochus ellipticus
size: 1 inch
range: whole coast

Zebra stylochus
Stylochus zebra
size: ½ inch
range: south of Cape Cod

Bdelloura candida lives comensally on the underside of the Horseshoe crab, *Limulus polyphemus*. Apparently it does not harm the Horseshoe crab in any way. It may merely share in the food of its host. *Bdelloura* attaches to the Horseshoe crab with an adhesive plate at its posterior end.

The Oyster flatworm, *Stylochus ellipticus*, is pale (as are most flatworms), with a band of eyespots along the front margin and a pair of tiny tentacles. It glides into oysters to eat them.

The Zebra stylochus, *Stylochus zebra*, is dark with light crossbands and is often found in hermit crab shells.

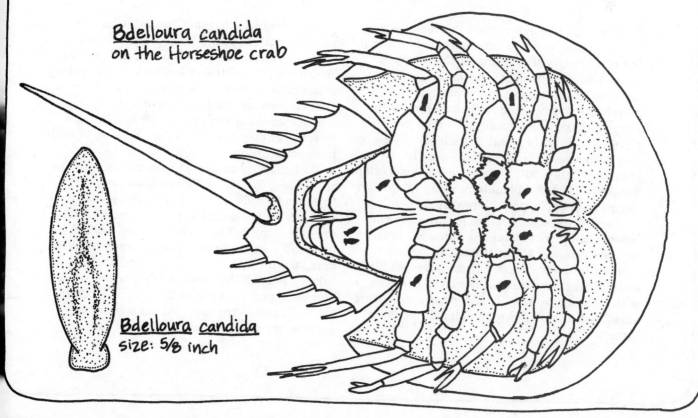

Bdelloura candida
on the Horseshoe crab

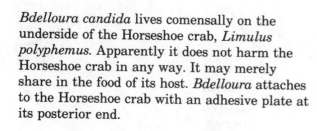

Bdelloura candida
size: 5/8 inch

Worms in the Plankton

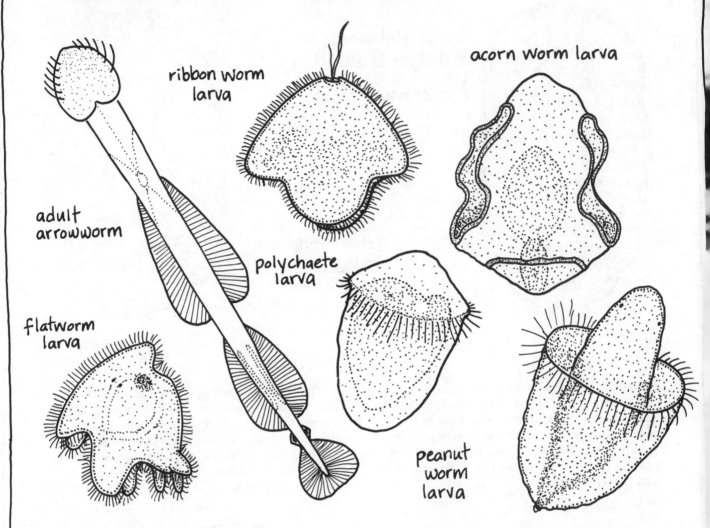

adult arrowworm

ribbon worm larva

acorn worm larva

polychaete larva

flatworm larva

peanut worm larva

Most worms, like many other marine invertebrates, spend at least part of their life in the plankton, floating about at the mercy of waves and currents. Ribbon worms, acorn worms, flatworms, peanut worms, and the polychaete worms are usually members of the *meroplankton*, which means that they are planktonic for only a part of their life. The eggs and/or larvae of the worms are usually planktonic; the adult worms are benthic (bottom dwellers). Having a planktonic stage allows the worms to disperse over a large area.

Not every species in these groups of worms belongs to the meroplankton. Some lay eggs that are attached to the bottom, rather than floating in the plankton. Others have what is called direct development: the larva develops within the egg so that it hatches as a miniature adult without going through planktonic larval stages.

Arrowworms (phylum Chaetognatha) are planktonic their entire lives, thus they are considered holoplankton. Chaetognath means "bristly jaw," and this describes the stiff hairs around the arrowworm's mouth. Most arrowworms are between one and three centimeters long and are common in inshore waters, yet they are hard to see because of their transparent body.

Without a microscope, arrowworms look like slivers of glass. But with magnification, their fins, head, trunk, and tail are readily visible. Eyes are on the dorsal surfaces. Arrowworms are carnivorous. The curved bristles around the head are used to capture prey. Arrowworms dart forward (like an arrow), spread their bristles to seize the prey, which is usually other plankton, and then pierce the prey with their teeth. *Sagitta* is the most common genus of arrowworm on our coast.

True Segmented Worms

By far, the most common marine worms belong to the phylum Annelida, the true segmented worms. For some reason, most beachcombers are about as thrilled to see a worm as they are to see a Great white shark. Because of this unfortunate aversion towards these innocuous invertebrates, few people realize just how ubiquitous annelid worms are.

Picture yourself sitting at the shore. Unbeknown to you, hundreds of annelids are in the sand, under the seaweed, beneath the rocks, attached to pilings, and swimming in the water. Yet, because of the secretive, generally inconspicuous nature of marine worms, you continue to enjoy the beach, blissfully unaware of the wriggling biomass around you. Little do you realize that worms can be fun, and interesting!

Annelids are distinguished from other worm-like invertebrates by the rings around their bodies, which divide their bodies into segments. (An earthworm is an example of an annelid.) Actually, only the trunk is segmented. Neither the head, which contains the brain, nor the terminal end, containing the anus, is a true segment. Each segment bears identical lateral nerves, blood vessels, and excretory organs. As the worm grows, new segments are added between the trunk and the terminal end. The oldest segments are found near the head; anterior trunk segments often fuse with the head.

Waves of muscular contractions that pass down the worm's body elongate and contract segments, allowing the worm to move ahead. Small paired bristles on each segment help increase traction.

There are three classes of annelids: Hirundinea, the leeches; Oligochaeta, the "earthworms"; and Polychaeta, the class containing the vast majority of marine species. Only the polychaetes will be discussed here.

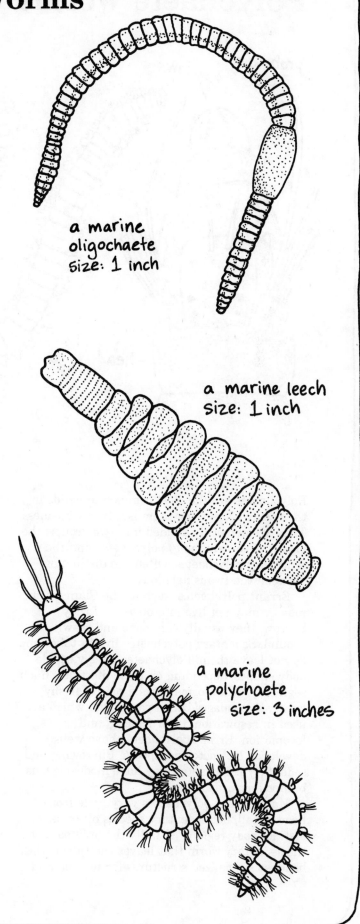

a marine
oligochaete
size: 1 inch

a marine leech
size: 1 inch

a marine
polychaete
size: 3 inches

Polychaete Worms

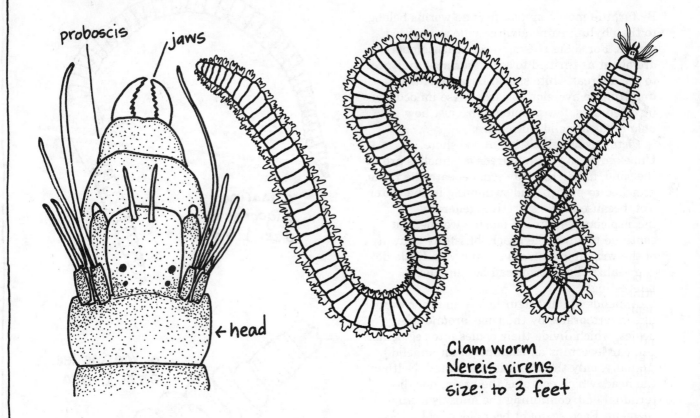

proboscis

jaws

← head

Clam worm
Nereis virens
size: to 3 feet

Polychaetes are the most common annelids in the ocean. Classifying them is a tricky business. They are generally divided into two groups: the errant (free-swimming) polychaetes and the sedentary polychaetes. Often the distinction between the two is nebulous.

Errant polychaetes, such as the Clam worm, may or may not live in a burrow. If they burrow, they usually can leave and crawl around. Sedentary polychaetes live in tubes and do not leave them. Polychaetes have variable body shapes, depending on their lifestyles. Each segment of the worms bears a pair of fleshy bristly appendages called parapodia, which are used in respiration, filter feeding, and locomotion. Errant polychaetes have well-developed locomotor parapodia. Burrowing and sedentary polychaetes may have inconspicuous bumps for parapodia.

Many errant polychaetes vary little from head to tail, but sedentary species often have distinct body regions. Usually the only part of the sedentary worm that leaves the tube is the head, so specialized structures for feeding and respiration are found there. Most sedentary polychaetes feed on plankton or detrital particles that fall on or get trapped in the feeding tentacles. On the other hand, errant polychaetes are usually carnivorous and have specialized structures for prey capture, such as strong jaws and/or a proboscis. Some polychaetes ingest the mud through which they burrow.

Sexes are usually separate in polychaetes: a worm is either a male or a female. Typically, egg and sperm are released into the water. After fertilization, planktonic larvae develop and then metamorphose into adults. Most polychaetes are fragile, but many can regenerate lost parts.

Clam worms, _Nereis virens,_ live intertidally and subtidally from Maine to Delaware. During the daytime they stay in mucus-lined sand burrows, but at night they come out to prey on small crustaceans and mollusks. When a clam worm detects food, it quickly thrusts its proboscis outward and seizes the prey with its formidable jaws.

Bristle and Paddle Worms

Green bristle worm
<u>Hermodice</u> <u>carunculata</u>
size: 10 inches

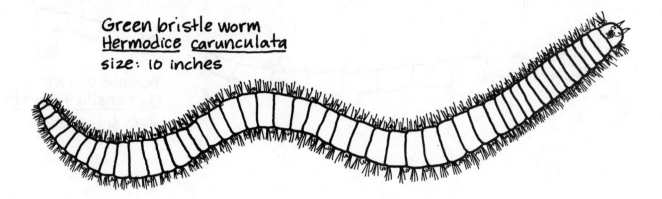

As beautiful as fire worms are, their bent for "stinging" people and eating coral has given them a bad name. Fire worms, also called bristle worms, are found in southern Florida and the Caribbean on reefs, under stones, and in grassy or sandy areas. When handled, their hollow bristles break off and easily penetrate human flesh. A painful irritation results; thus the name fire worm.

The Green fire worm, *Hermodice carunculata,* feeds on corals and sea anemones. It opens its mouth, engulfs a branch of staghorn coral, and stays there five to ten minutes digesting the coral polyps and the tissue connecting them. Only the naked branch tips of the coral skeleton remain. Green fire worms are bright green with red gills and white bristles.

Another errant polychaete is the paddle worm, *Phyllodoce arenae,* named for the paddle-shaped flaps on its parapodia. Paddle worms are found along the entire coast, creeping among the flora and fauna on the bottom. They are known to fraternize with sponges and barnacles.

Like most other errant polychaetes, paddle worms are very active carnivores. Although their proboscis does not have jaws, that does not diminish the worm's ability to zap unsuspecting prey. Paddle worms have simple eyes that are sensitive to light intensity and direction but do not discern actual images. *Phyllodoce arenae,* only one of many species of paddle worm, lives from Maine to North Carolina.

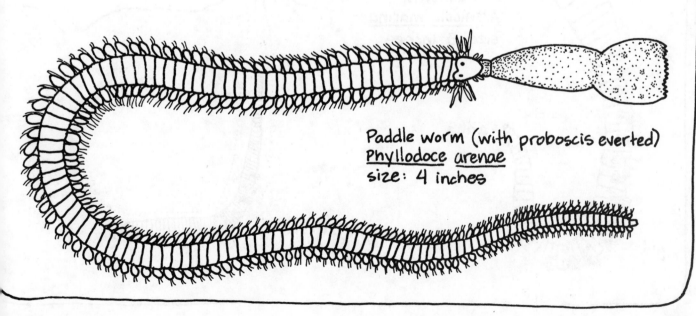

Paddle worm (with proboscis everted)
<u>Phyllodoce</u> <u>arenae</u>
size: 4 inches

Bamboo Worms and Lugworms

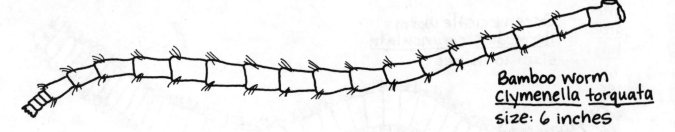

Bamboo worm
<u>Clymenella</u> <u>torquata</u>
size: 6 inches

In the never-ending struggle to give organisms logical common names, the Bamboo worm, *Clymenella torquata*, stands out as a shining success. Its parapodia are reduced to small ridges, giving this worm the jointed look of a bamboo cane. To complete the bamboo motif, the worm's head is blunt without any appendages.

Bamboo worms live upside down in the substrate in a tube of sand grains. They ingest sand and mud as they burrow. Food particles are digested and extra dirt is excreted out of the burrow after it passes through the worm's gut. Bamboo worms are found from Maine to North Carolina, mainly subtidally.

Like many sedentary polychaetes, the Lugworm, *Arenicola*, is divided into distinct regions: the head (without parapodia), the trunk (with gills and parapodia) and the tail (without any appendages). Lugworms excavate an L-shaped burrow in the substrate, ingesting sand along the way with their simple proboscis. They back up to the top of the burrow to discharge the semi-digested dirt. It is thought that Lugworms also feed on tiny particles brought in with water currents that they pump through their burrow. Special balancing organs called statocysts keep the worm burrowing downward. *Arenicola marina* is found north of Cape Cod; *Arenicola cristata* is found from Cape Cod to the Caribbean.

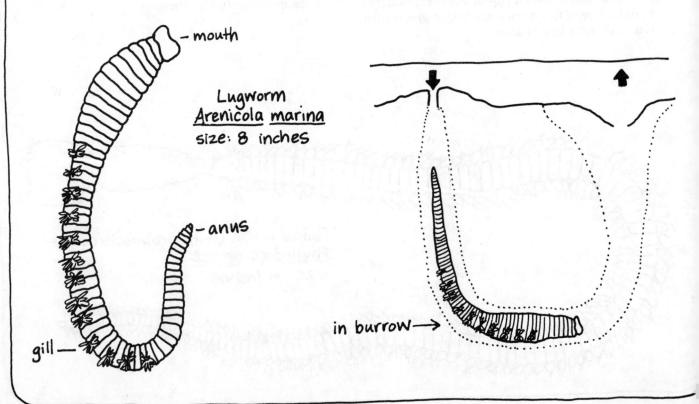

– mouth

Lugworm
<u>Arenicola</u> <u>marina</u>
size: 8 inches

– anus

gill –

in burrow →

Scaleworms

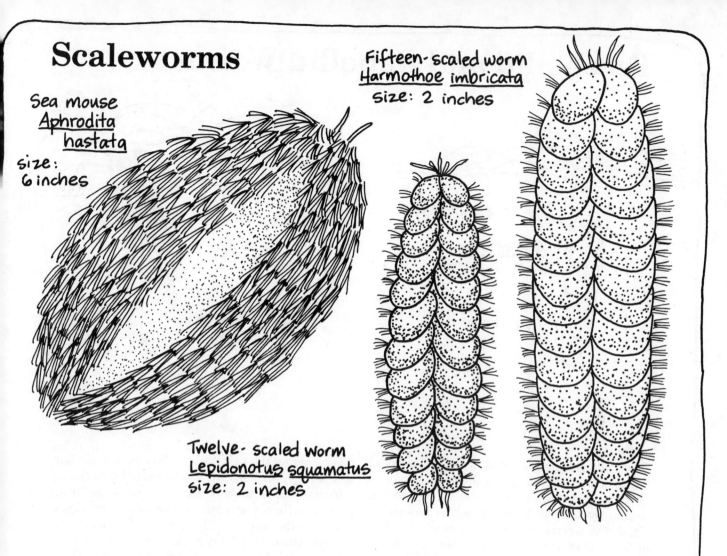

Sea mouse
*Aphrodita
hastata*

size:
6 inches

Fifteen-scaled worm
Harmothoe imbricata
size: 2 inches

Twelve-scaled worm
Lepidonotus squamatus
size: 2 inches

Not all errant polychaetes look as obviously segmented as a clam or fire worm. Consider the scale worms. At first glance they resemble pill bugs or armadillos; they barely look like worms. Overlapping plate-like scales on the dorsal side of scale worms effectively disguise their segmented look, although the segments are clearly visible from underneath. Each scale is attached to the body by a small stalk.

Two of the more common scale worms are *Lepidonotus squamatos,* the Twelve-scaled worm, and *Harmothoe imbricata,* the Fifteen-scaled worm. As their names imply, the Twelve-scaled worm has twelve (or thirteen!) pairs of scales; the Fifteen-scaled worm has fifteen pairs. Both species are found from Maine to Chesapeake Bay. Other scale worms range as far south as the Caribbean.

When they are disturbed, scale worms may roll up in a ball or shed scales, which they can later regenerate. Finding scale worms is not difficult. They are very common intertidally, under rocks and in tidepools, where they forage for food. Another unique feature about these atypical polychaetes is their eggs. Instead of releasing their eggs into the water, scale worms brood them.

Even more anomalous than the normal, everyday scale worm is the Sea mouse, *Aphrodita hastata.* The entire dorsal surface of this rather flashy scale worm is covered with hair-like felt, which is often iridescent. Beneath this hirsute exterior lie fifteen pairs of scales. Sea mice are found from Maine to New Jersey. They live subtidally and are seldom seen unless washed ashore after a storm.

Terebellid and Sabellid Worms

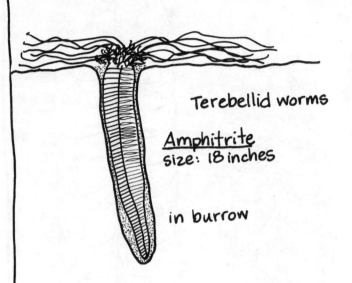

Terebellid worms

<u>Amphitrite</u>
size: 18 inches

in burrow

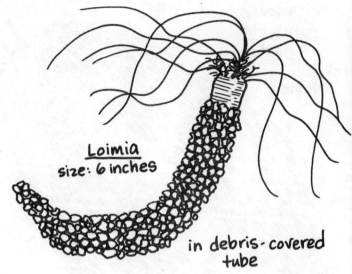

<u>Loimia</u>
size: 6 inches

in debris-covered
tube

Terebellid worms are sedentary tube-dwellers, living either in a burrow in the sand or in a particle-encrusted tube under rocks and shells. Sedentary does not imply immobile, however. Terebellid worms can pull their head in and out of their tubes, exposing their gills and tentacles.

Terebellids stretch their feeding tentacles out over the substrate. Food particles in the water float down and stick to a mucus adhesive secreted by the tentacles. Ciliary action, similar to a conveyor belt, moves the particles to the base of the tentacles where they are wiped into the mouth. Some tentacles extend several times beyond the worm's body length. When the worm is disturbed, it quickly pulls its tentacles back into its tube. This action looks very much like someone sucking in strands of spaghetti.

Fan worm
<u>Sabella micropthalma</u>
size: 1½ inch

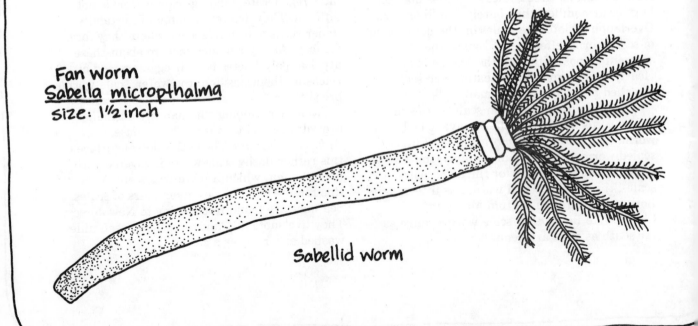

Sabellid worm

Trumpet and Parchment Worms

The Trumpet worm, *Pectinaria gouldi*, called the Ice cream cone worm by dairy enthusiasts, is an attractive but secretive tube-dwelling worm. It buries itself upside down within a cone-shaped tube. The tube, an amazing construction feat, is made of a single layer of quartz sand grains glued together. Empty cones can be found on the beach.

With its tentacles, the Trumpet worm digs head first through the substrate, ingesting sediment. Food particles are digested and the dirt is ejected to the surface through the end of the tube. The Trumpet worm lives subtidally and intertidally along the entire coast.

Parchment worms, *Chaetopterus veripedatus*, live in U-shaped burrows open at both ends. Small piles of dirt mark the openings. These unique worms use three paddles on their middle region to create a water current through the tube. Plankton is filtered out with a mucus bag near the paddles. Periodically, a special organ removes the bag and sends it to the worm's mouth along a ciliated groove. Parchment worms are chiefly subtidal and live from Cape Cod to the Caribbean.

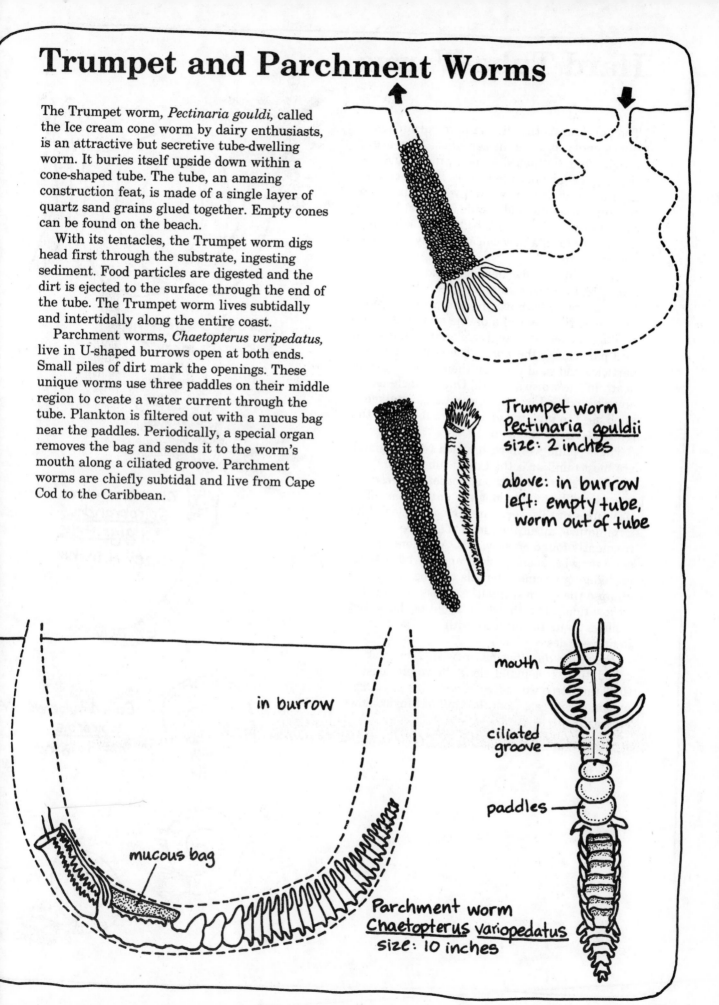

Trumpet worm
Pectinaria gouldii
size: 2 inches

above: in burrow
left: empty tube,
worm out of tube

in burrow

mucous bag

mouth

ciliated groove

paddles

Parchment worm
Chaetopterus variopedatus
size: 10 inches

Hard Tube Worms

The only worms that live in hard calcareous (limestone) tubes are the serpulids. They are usually found attached to algae, rocks, and shells. A collar near the worm's head secretes the tube in which the worm lives. A mixture of calcium carbonate and mucus flows out between the collar and the worm's body. This secretion hardens to form a new calcareous layer at the end of the tube.

Like most tube-dwellers, serpulids are highly specialized for their sedentary life. Their tentacles serve both feeding and respiratory functions. Hair-like cilia on the tentacles beat to produce a current, bringing plankton and small food particles to the worm. The tentacles trap particles and send them on their way to certain death in the worm's mouth. One tentacle is always modified into an operculum, a stalked knob used to plug the end of the tube when the tentacles are withdrawn.

The serpulid worm with which New Englanders are most familiar is the Coiled tube worm, *Spirorbis spirillum*, although other *Spirorbis* worms can be found along the entire coast. The tiny tubes of this worm are coiled and look like a miniature, albeit flat, snail. They are frequently found on seaweeds. A patient observer with keen eyesight and a prediliction for hanging around tidepools may be able to glimpse the worm while it extends its tentacles. At low tide, when the worm is out of the water, it plugs up its tube tightly with the operculum to prevent desiccation.

Hydroides dianthus has a twisted, but not coiled, tube. It is found along the whole coast. Christmas tree worms, *Spirobranchus giganteus*, live on coral reefs from the Gulf of Mexico down to Brazil.

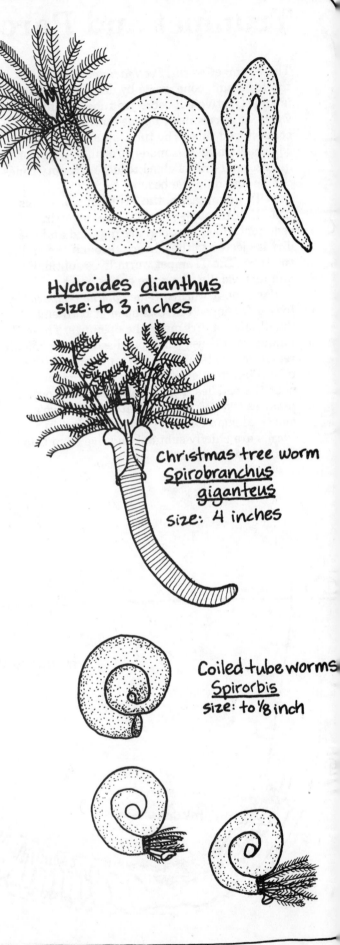

Hydroides dianthus
size: to 3 inches

Christmas tree worm
Spirobranchus giganteus
size: 4 inches

Coiled tube worms
Spirorbis
size: to 1/8 inch

Marine Worm True/False Quiz

1. Sea mice are blind.

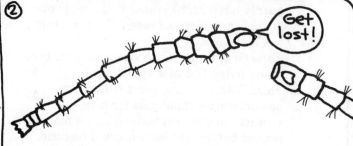

2. Bamboo worms are very territorial and are rarely found within six inches of each other.

3. Although peanut worms are not as tasty as peanuts, Norwegians grind them into an edible paste.

4. Acorn worms are brightly colored.

5. Some flatworms can sting.

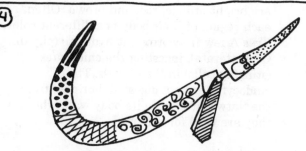

6. Clam worms make great fish bait.

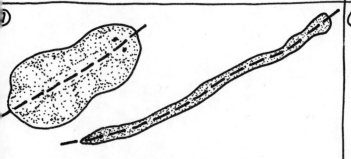

7. Marine worms are bilaterally symmetrical.

8. It is easy to photograph the tentacles of fan worms.

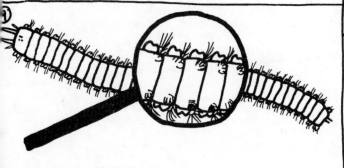

9. All marine worms have gills.

10. Most marine leeches suck blood.

1. *False* Sea mice are not blind, although they hardly have 20/20 vision. Two simple eyes at their anterior end detect light intensity, but little else.
2. *False* Gregarious Bamboo worms have been found living 500 to a square yard.
3. *False* With their introvert withdrawn, peanut worms could pass for a shelled peanut. No one grinds them into a marine peanut butter, although it could be done.
4. *True* Not all marine worms are as dull as their terrestrial counterparts. Acorn worms are bright orange, red, and brown. Often, each region of their body is a different color.
5. *True* A few flatworms eat hydroids (phylum Coelenterata), ingesting the cnidocytes (stinging cells) in the process. The cnidocytes are not digested, but pass from the flatworm's gut to its body wall, where they are used for defense.
6. *True* Clam worms (*Nereis*) and bloodworms (*Glycera*) are commonly sold for bait. In the wild, both these worms are eaten by crustaceans, birds, and bottom-dwelling fish.
7. *True* Free-moving animals, like worms, generally are bilaterally symmetrical. Sessile animals (those attached to the bottom) are often radially symmetrical.
8. *False* Fan worms are very sensitive to vibrations and light, and withdraw their tentacles at the drop of a hat.
9. *False* Not all worms need structures as complex as gills to obtain oxygen. For example, flatworms and ribbon worms take in oxygen through their body wall.
10. *True* Seventy-five percent of known species are blood-sucking parasites that attach to snails, worms, crustaceans, and fish. Other leeches are predators.

9 Bryozoans

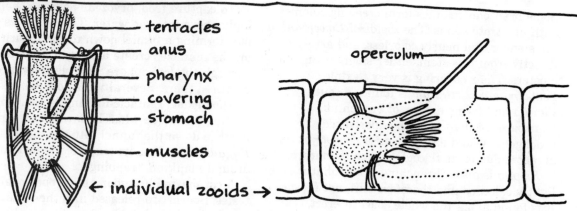

tentacles
anus
pharynx
covering
stomach
muscles

← individual zooids →

operculum

Colonies of bryozoans (phylum Bryozoa) look even less like animals than do the sponges, if that is possible. Although bryozoans live abundantly in coastal water on pilings, shells, rocks, and algae, people often mistake them for seaweed, moss, or water-logged spaghetti.

An individual bryozoan animal is called a zooid, but since bryozoans are colonial, zooids are never found alone. Zooids vaguely resemble hydroids, but are much more complex anatomically. They are minute (less than one thirty-second of an inch), and come in a variety of shapes, including oval, box, vase-like, and tubular. Encircling the zooid's mouth is a specialized food-catching organ, the lophophore, an **O**- or **U**-shaped fold in the body wall. Numerous tentacles branch off the lophophore. The zooid's anus opens through the lophophore to the outside. Another name for the phylum Bryozoa is the phylum Ectoprocta, literally meaning "outside anus." Periodically, the lophophore and gut degenerate, and are later regenerated by the zooid.

Bryozoan zooids live together in sessile (attached) colonies. Colonies can be branching or encrusting. Most are enclosed in a protective covering secreted by the zooids themselves. Calcium carbonate (limestone) is the most prevalent type of covering. Some colonies are only lightly calcified, making them slightly stiff. Other colonies are heavily calcified, resulting in a hard, crunchy covering. A few bryozoans secrete a flexible protein cuticle instead. Regardless of the type of covering, the only part of the zooid that protrudes above the covering is the lophophore. It can be extended into the water through a hole in the covering. Many bryozoans have an operculum (lid) over this opening.

Encrusting bryozoans are the most common type. Zooids form a sheet-like colony attached to hard objects and usually the colony has a hard calcareous covering. *Membranipora* forms a delicate white lacy crust on rocks and seaweed along the entire coast. *Watersipora* forms an orangy crust in the Caribbean and south Florida.

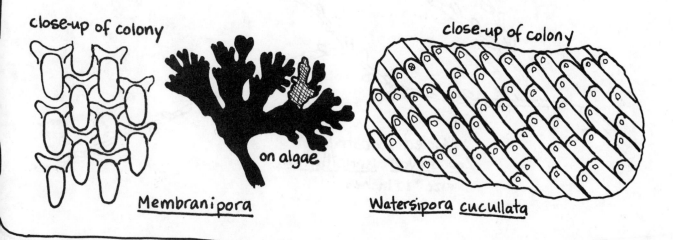

close-up of colony

on algae

Membranipora

close-up of colony

Watersipora cucullata

Bryozoans

Some bryozoan colonies form creeping colonies with prostrate stems. The zooids of *Bowerbankia* are slender and nearly colorless, and arise directly from this stem, either singly or in clusters. Their covering is very flexible. *Zoobotryon verticillatum* is found in south Florida and forms masses of branching, transparent colonies. Even experienced beachgoers would easily dismiss *Zoobotryon* as an algae due to its flaccid, branching nature.

Bugula is also a branching colony, but it is lightly calcified and therefore stiff, not flaccid. It is found along the whole coast in erect, tufted, plant-like colonies. The zooids are arranged in double rows on the branches.

To capture food, bryozoans extend their lophophore into the water. The tentacles fan out, forming a funnel down to the mouth. Cilia on the tentacles create a current of water through the lophophore, and plankton are captured there. Several variations on this method exist. Tentacle-flicking is one. This is a feeding technique whereby a tentacle pushes a particle down the funnel into the mouth. *Bugula nerita* uses its tentacles to form a cage around plankton, trapping it.

Most marine bryozoans have both ovaries and testes. Sperm are released into the water and caught by the tentacles of another zooid. After fertilization some species release their eggs into the water, but most bryozoans brood their eggs a while.

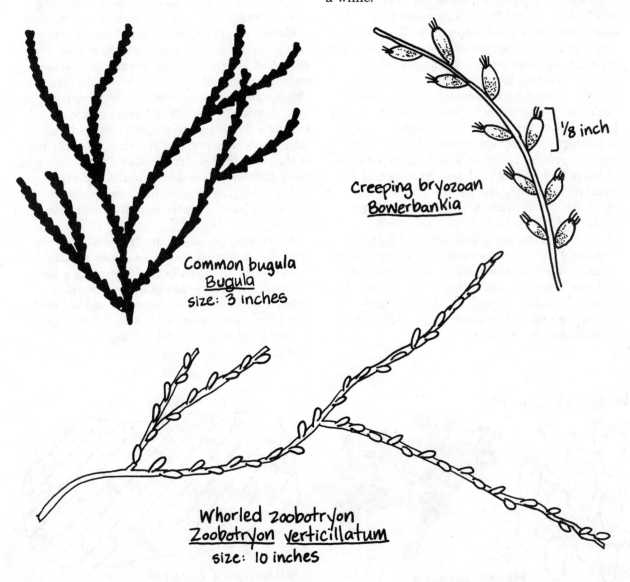

Common bugula
<u>Bugula</u>
size: 3 inches

Creeping bryozoan
<u>Bowerbankia</u>

1/8 inch

Whorled zoobotryon
<u>Zoobotryon verticillatum</u>
size: 10 inches

Bryozoan True/False Quiz

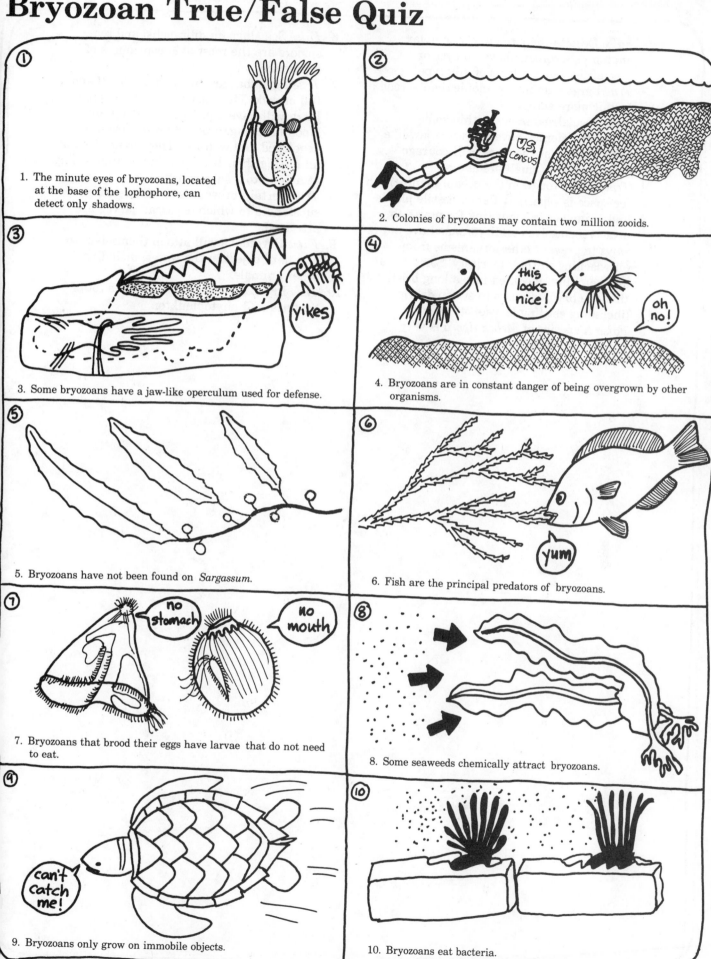

1. The minute eyes of bryozoans, located at the base of the lophophore, can detect only shadows.

2. Colonies of bryozoans may contain two million zooids.

3. Some bryozoans have a jaw-like operculum used for defense.

yikes

4. Bryozoans are in constant danger of being overgrown by other organisms.

this looks nice!

oh no!

5. Bryozoans have not been found on *Sargassum*.

6. Fish are the principal predators of bryozoans.

yum

7. Bryozoans that brood their eggs have larvae that do not need to eat.

no stomach

no mouth

8. Some seaweeds chemically attract bryozoans.

9. Bryozoans only grow on immobile objects.

can't catch me!

10. Bryozoans eat bacteria.

1. *False* Despite their relatively complex anatomy, bryozoans have no eyes or specialized sense organs of any type.
2. *True* Large encrusting colonies can spread a foot or more across.
3. *True* Specialized zooids within some bryozoan colonies have evolved a jaw-like operculum that they use to discourage and/or destroy small predators.
4. *True* Life is tough in the ocean. Competition for space is especially fierce. Sessile plants and animals are competing for places to attach themselves to. Bryozoans have many ways to prevent other organisms from encroaching in their territory. Some have an operculum modified into a long bristle that the bryozoan uses to sweep debris (including settling larvae) off itself.
5. *False* A species of *Mebranipora* is very common on *Sargassum*.
6. *False* Sea slugs (nudibranchs) and sea spiders are the most avid consumers of bryozoans.
7. *True* Eggs that are brooded are large and full of yolk. This ensures the developing larva will have a plentiful food supply. Upon hatching, brooded larvae are well developed and well fed. They do not feed at all during their brief planktonic fling before attaching to a substrate.
8. *True* Certain brown algae may release substances to which bryozoan larvae are attracted.
9. *False* Bryozoans will attach themselves to any hard object in the water, including other animals.
10. *True* Bryozoans seem to consume a lot of bacteria and phytoplankton.

10 Mollusks

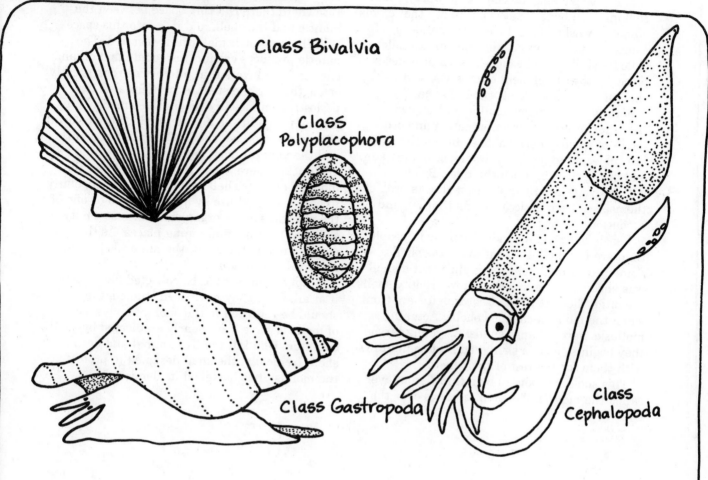

Class Bivalvia

Class Polyplacophora

Class Gastropoda

Class Cephalopoda

Beachgoers unintentionally ostracize or ignore many invertebrates along the shore. Mollusks, however, seldom go unnoticed, because of their showy shells. People who wouldn't give the time of day to a worm will fall all over a snail or a clam. Such is life at the seashore.

The phylum Mollusca contains 100,000 species divided into seven classes, four of which are found close to shore: class Gastropoda, the snails; class Polyplacophora, the chitons; class Bivalvia, the two-shelled (clam-like) mollusks; and the class Cephalopoda, the squids and octopi. Although mollusks are primarily marine, there are freshwater and terrestrial species.

Despite the apparent lack of similarities between these classes, all mollusks have several features in common. Their body is divided into three regions: the head, the foot, and the visceral mass. The head contains the mouth and sensory organs in all classes but the bivalves, in which the head is indistinct. The fleshy foot is used for crawling, swimming, or burrowing. Organs of respiration, circulation, reproduction, digestion, and excretion are located in the visceral mass—the main part of the body. These organs include a heart, a stomach, intestines, gonads, and kidneys.

A Shell Is a Skeleton

While eating clams, no one has to worry about choking on a bone. That is not to say that clams do not have skeletons. Like most mollusks, clams have a calcareous (limestone) exoskeleton called a **shell.** Just as our skeleton provides a place for the attachment of muscles, so does the shell of a mollusk. Being soft-bodied and generally slow-moving, mollusks also need shells for protection from predation and other forces of nature. Periwinkle snails on the shore are subjected to pounding waves, scorching sun, and dry periods between the tides. By withdrawing into their shells, mollusks protect themselves from desiccation, mutilation, and predation.

Seashells are much more familiar to people than are the mollusks that made the shells. Some people may not even realize that a shell was once part of a living creature. Empty shells are often thought to be discards, leftovers from when the mollusk changed shells. Actually, mollusks never change shells. As young larvae they begin to form a shell, and that shell grows with them for the rest of their lives.

This shell is produced by a thin tissue layer called the *mantle*. The mantle hangs over the body, touching the shell in only a few places. Elsewhere there is a tiny space between the mantle and the shell, and it is into this space that the new shell material is secreted. The mantle produces calcium carbonate and organic chemicals in layers that often crisscross, greatly strengthening the shell. A protein layer (the periostracum) usually covers the outside. Pigment cells in the mantle produce the colorful patterns of mollusk shells. Beneath the mantle and the visceral mass is a cavity containing gills and ducts, where waste is expelled. When the mollusk dies, its shell remains behind, hopefully to be put to good use by a hermit crab or one of the countless other organisms that use empty shells as habitats and hiding places. As the empty shell deteriorates, the minerals in it are returned to the sea.

Empty shells should be collected judiciously, if at all. Once they have been studied, they should be returned to the shore. Many species of mollusk have been nearly eradicated by shell collectors who kill the live animal and keep its shell. Ironically, the shell designed to protect the mollusk has hastened its demise.

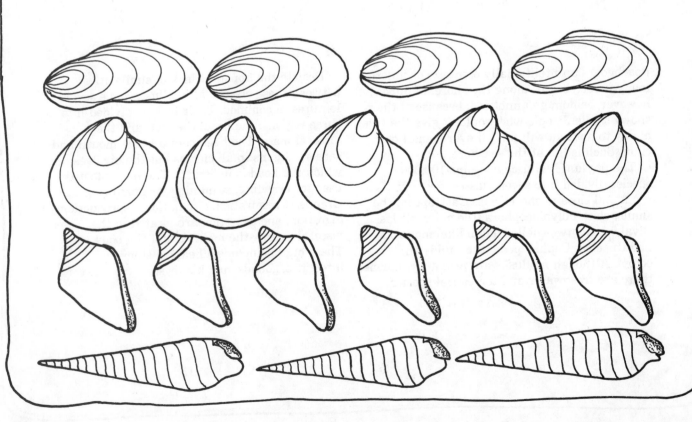

Feeding

All mollusks except the bivalves have a feeding organ called the radula. The radula is a tongue-like structure covered with fine teeth. Depending on the feeding habits of the mollusk, the radula is adapted for grating, scraping, grasping, or cutting. *Littorina littorea,* an intertidal periwinkle snail, uses its radula to scrape minute bits of algae off rocks. Mucus on the radula entangles food particles.

Many carnivorous snails, such as the Northern whelk, *Buccinum undatum,* have an extensible proboscis. This tube-like organ contains the radula, mouth, and esophagus. While holding a bivalve in a death grip with its foot, the Northern whelk wedges apart the two shells of the bivalve with the edge of its own shell. Once the shell is opened, the whelk inserts its proboscis and eats the bivalve's body.

Most bivalves are filter or suspension feeders. The Rigid pen shell, *Atrina rigida,* which lives partially buried in the bottom, is an example. Cilia on its gills create a current, bringing water and the plankton suspended in it into the partially opened shell. The gills filter out the plankton, which becomes trapped in mucus and is transported to the mouth.

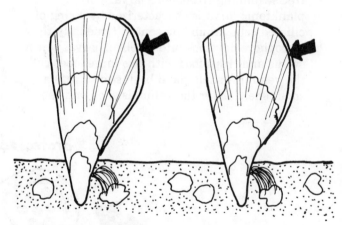

Cephalopods also have a radula, but their primary feeding structure is their beak. Squid swim after their prey and use their two longest arms to seize their victims, usually fish or smaller squid. The other eight arms help to hold down the prey and draw it toward the mouth. Then the squid uses its strong parrot-like beak to tear chunks of flesh out of its prey.

Mollusk Life History

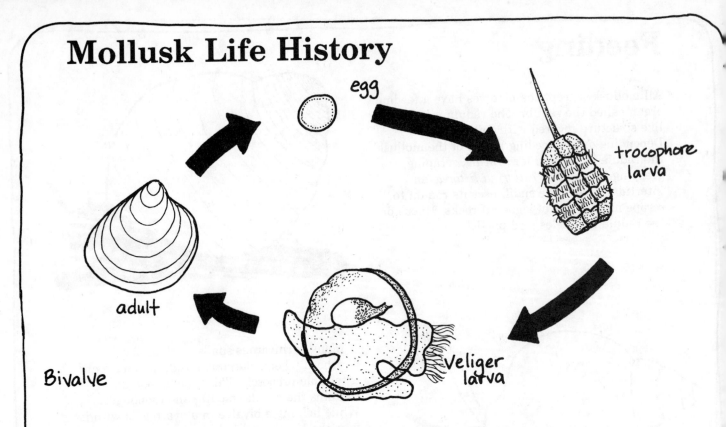

Most mollusks are dioecious; males and females are separate individuals. Bivalves typically release their sperm and egg cells directly into the water, where fertilization then occurs. The planktonic fertilized egg develops first into a free-swimming trocophore larva. This planktonic larva is characterized by a ring of ciliated cells around its center. Gradually the trocophore develops into a veliger larva, which swims and traps food with two large lobes called the velum. A shell gland begins to secrete a small shell. Later the velum is shed and the bivalve larva settles to the bottom, where it grows into an adult, if all goes well.

Most gastropods are also dioecious, but fertilization is internal. The fertilized eggs are generally deposited in a protective ribbon, string, case, or capsule that lies on or is attached to the bottom. Usually the trocophore stage occurs within the egg, so upon hatching, a swimming veliger larva emerges. As a foot develops, the larva settles to the bottom. Some gastropods emerge from their egg case as fully developed miniature snails.

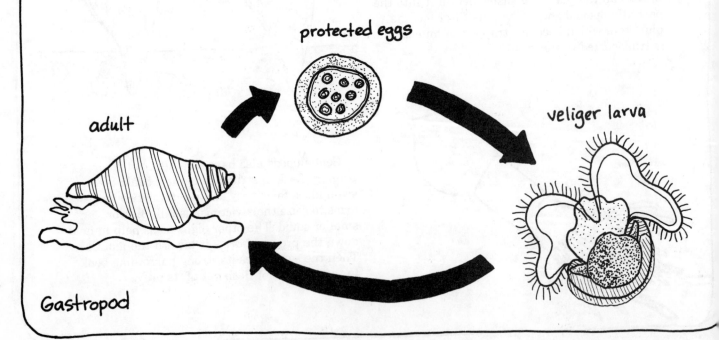

Gastropods

With 75,000 species, gastropods (literally "stomach foot") comprise the largest class of mollusks. Another term for this class is univalve, meaning "one-shell." With apologies to malacologists, this class can be loosely referred to as the snails. Most gastropods have a well-developed head with tentacles and eyes at the base of the tentacles, and a radula. With a class this size, however, there are exceptions to every rule. Many gastropods have no shell (subclass Opistobranchia), but those that do, have one shell, usually spiral. The shells of some gastropods are hidden beneath their mantle.

Gastropods usually have a sole-like muscular foot adapted for crawling. Ciliated cells on the underside of the foot secrete a mucus trail over which the snail moves. Some gastropods, such as the worm shells, are sessile; they don't move at all. Although some gastropods live on hard bottoms, most live on sand or mud.

A snail's shell serves as a portable shelter. But snails are not loose inside their shell and cannot completely leave it. They have muscles attached to the columella, the central axis of the shell. The gastropod's head and foot extend out of the shell through the aperture, but can be completely withdrawn into the shell by a retractor muscle. The foot usually has a horny disc, the operculum, attached to it. When the snail pulls inside the shell, the operculum acts as a door, closing off the aperture and leaving the snail safely sequestered inside. Not only does this thwart some predators, but it helps prevent the snail from drying out should it be stranded high on the shore.

To keep a current of water circulating over their gills, many snails have an inhalant siphon, which is actually a rolled-up section of the mantle. The siphon may also have other uses, such as a sensory probe.

Gastropods have relatively well-developed sense organs. Most have simple eyes that are able to detect light intensity. The tentacles are used as feelers and chemoreceptors. Statocysts, equilibrium organs found in the foot, help the snail to orient with respect to the bottom. Patches of skin located on the snail's gill membranes are used to detect particles in the water passing over the gills.

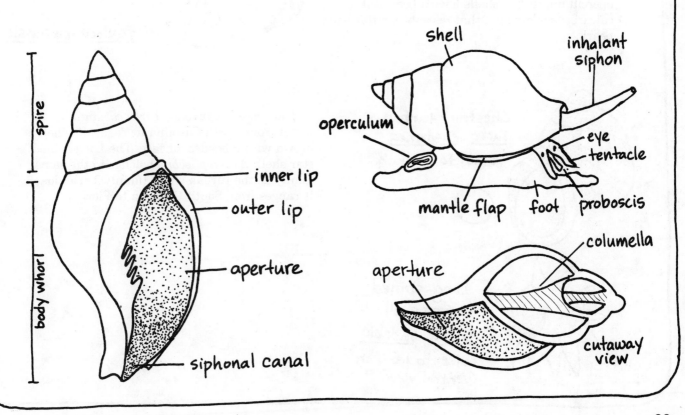

Limpets and Turban Shells

The true limpets are found intertidally on rocky shores around the world. Their flat, uncoiled shell, which resembles an Oriental straw hat, and their strong foot allow limpets to adhere tightly to rocks. Most limpets have a definite place of their own on a specific rock. If the rock is relatively soft, the limpet erodes it and creates a home spot that fits its shell perfectly.

At night, limpets wander around, scraping microscopic algae off rocks with their radula. After a trek of six to thirty-six inches, the limpet returns to its home spot. How it finds the spot is a bit of a mystery, since the limpet doesn't seem to retrace its route back home. Not all limpets have this homing instinct. The Atlantic plate limpet, *Acmaea testudinalis*, is a true limpet that lives intertidally from the Arctic to New York.

Keyhole limpets, such as *Diodora cayenensis*, which is found from New Jersey to Florida, have two gills, unlike the true limpets which have one. They are easily identified by the hole on top of their shell. Water is drawn in under the shell, passed over the gills, and secreted through the hole. Waste is also expelled there. *Diodora* browses around rocks at night.

Turban shells are herbivorous snails found in tropical and subtropical seas. They usually live in shallow water, where they feed on algae. Unlike most snails, which have a thin, dark operculum, turban shells have a beautiful calcareous operculum that resembles a flattened pearl.

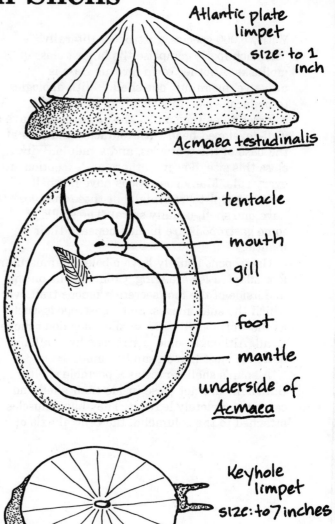

Atlantic plate limpet
size: to 1 inch
Acmaea testudinalis

tentacle
mouth
gill
foot
mantle
underside of *Acmaea*

Keyhole limpet
size: to 7 inches
Diodora cayensis

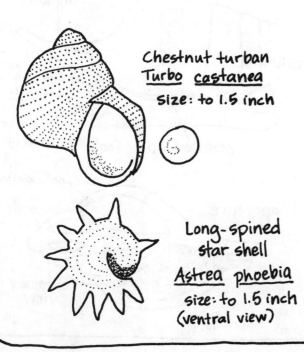

Chestnut turban
Turbo **castanea**
size: to 1.5 inch

Long-spined star shell
Astrea phoebia
size: to 1.5 inch
(ventral view)

The Chestnut turban, *Turbo castanea*, is found from North Carolina to Texas. Its shell is brown with a beaded surface. The Long-spined star shell, *Astrea phoebia*, belongs to the same family as the turban shells (family Turbinidae). It ranges from South Carolina to Florida.

Nerites and Periwinkles

Who would guess that under the drab brown shell of the Common periwinkle lurks meat over which Europeans drool? Yes, it's true, the Common periwinkle, *Littorina littorea,* found by the billions on intertidal rocks from Maine to Maryland, is considered a delicacy in Europe, where it is also common, but not as abundant.

Considering the huge populations of periwinkles on our shores today, it is hard to believe that these snails were not even discovered on the coast of North America until the mid 1800's. Whether they drifted over on logs from Europe, or were accidentally introduced, Common periwinkles quickly spread down the coast from Nova Scotia to Maryland.

Common periwinkles are herbivorous snails well adapted to the challenges of the rocky intertidal zone. Because the tide recedes twice a day, many of these gastropods are out of the water for hours at a time. To prevent desiccation, periwinkles seal themselves to rocks with mucus produced by their foot. While the mucus hardens, the snail pulls into its shell and closes its operculum. This creates a tight seal, preventing the periwinkle from drying out. When the tide comes back in, the periwinkles resume crawling over rocks, leaving a trail of mucus behind them.

A special adaptation of *Littorina littorea* is its ability to breathe air, instead of water, for short periods of time. Extra blood vessels in the mantle cavity permit the snail to extract oxygen from the air. It is believed that terrestrial snails evolved from ancient periwinkles.

Periwinkles eat algae they scrape off rocks with their radula. As the radula is pulled back into the snail's mouth, the radula folds lengthwise, and the algae particles mix with mucus, forming a gooey string that is pulled into the stomach. Despite the relatively small size of periwinkles, they gradually erode rock surfaces with their relentless scraping.

There are nineteen other species of periwinkles in North America. *Littorina irrorata,* the Marsh periwinkle, is common in marshy areas from New Jersey to Florida.

The tropical counterpart to the periwinkles are the nerites, common intertidal grazers in southern Florida and the Caribbean. Most have a toothed edge to their aperture. The Bleeding tooth, *Nerita peloronta,* has a bright blood-colored blotch along this edge.

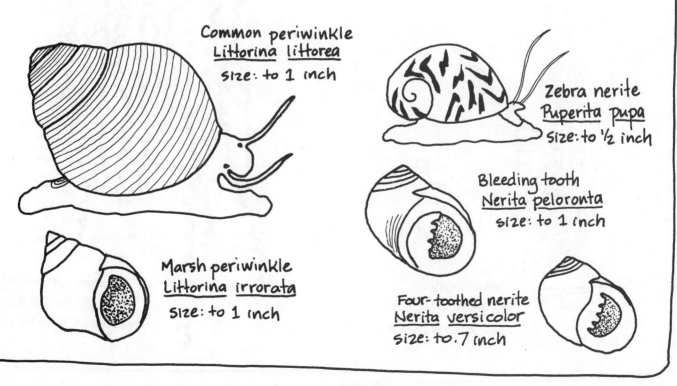

Common periwinkle
Littorina littorea
size: to 1 inch

Zebra nerite
Puperita pupa
size: to ½ inch

Bleeding tooth
Nerita peloronta
size: to 1 inch

Marsh periwinkle
Littorina irrorata
size: to 1 inch

Four-toothed nerite
Nerita versicolor
size: to .7 inch

Worm Shells and Janthina Shells

Not all snails pass the time crawling. The Janthina snail, *Janthina janthina*, spends its days floating attached to and beneath a raft of gelatinous bubbles secreted by its foot. Its purple shell is very fragile, and an operculum is present only in the larva.

The rather delicate look of *Janthina* gives no hint of its fearless eating habits. *Janthina* temporarily abandons its bubble raft to feed on the Portuguese man-of-war, *Physalia physalia*, and its close but innocuous relative, the By-the-wind-sailor, *Velella velella*. After storms, *Janthina* shells are frequently blown onto the shore. Live snails are found floating in warm waters on both coasts of the United States.

Worm shells, *Vermicularia*, also lead an atypical life for a gastropod. As larvae and juveniles, they look like normal snails, but as they mature, the whorls of their shells separate. Adult worm snails are more or less corkscrew-shaped. They live attached to rocks, shells, and sponges. Desite their odd shell, the snail is relatively normal-looking, except for its reduced foot.

Worm shells usually trap food on their gills, but some species secrete mucus threads near their aperture to trap food. Worm shells are found from Massachusetts to Florida.

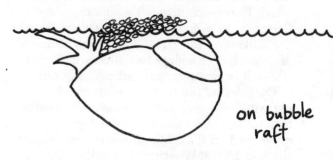

on bubble raft

Janthina janthina
Common janthina
size: to 1 inch

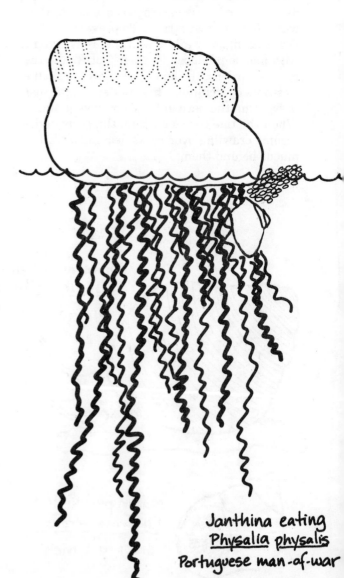

Janthina eating
Physalia physalis
Portuguese man-of-war

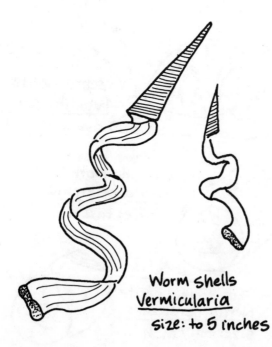

Worm shells
Vermicularia
size: to 5 inches

102

Slipper Shells

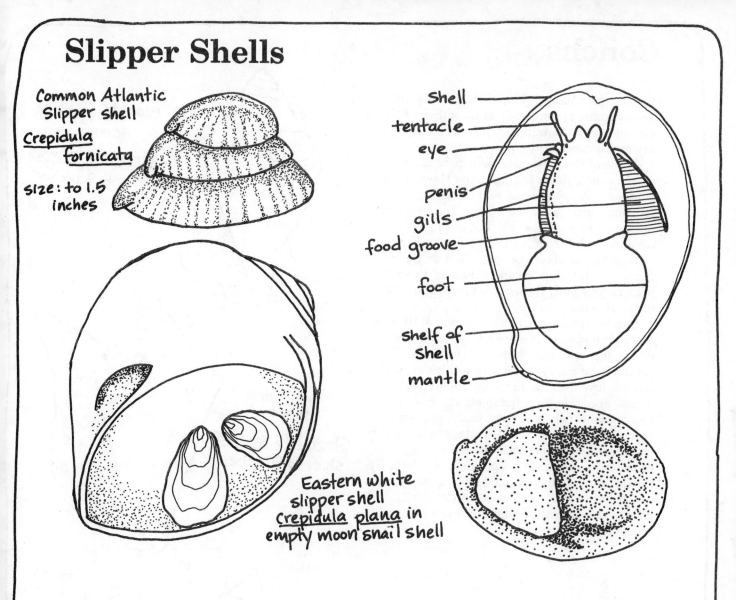

Common Atlantic Slipper shell
Crepidula *fornicata*

size: to 1.5 inches

Shell
tentacle
eye
penis
gills
food groove
foot
shelf of shell
mantle

Eastern white slipper shell *Crepidula* *plana* in empty moon snail shell

Slipper shells are snails that look like half a bivalve. A shell platform extending halfway across the underside of their shelf gives them the appearance of a boat or slipper. This shelf functions as support for their visceral mass. The coiled nature of their shell has disappeared, and the snail lives within an enlarged body whorl.

Although they have a foot on which they can move, most slipper shells are sessile. *Crepidula plana,* the Eastern white slipper shell, lives attached to the inside of empty snail shells. *Crepidula fornicata,* the Common Atlantic slipper shell, clings to rocks in the intertidal zone. Both species are found from Maine to Florida.

Slipper shells feed on particles suspended in the water. Their shell and mantle are held tightly against the substrate, except for a slight gap on each side of the shell. Water enters through the left side and leaves through the right. As plankton passes through, it is trapped in the mucus-coated gills. Later the snail uses its radula to remove the food and bring it to its mouth.

Slipper shells have rather anomalous sexual arrangements. They are often found in stacks, with older, larger individuals on the bottom, and smaller, younger ones on top. Young slipper shells are always male. Depending on the sex ratio of the surrounding individuals, the male reproductive tract may degenerate, allowing female gonads to develop.

Conchs

Conchs (pronounced "konks") are snails that have become synonymous with southern Florida and the Caribbean. In fact, natives of Key West are called conchs.

Three of the most common conchs are the Queen conch, *Strombus gigas*, the Hawk wing conch, *Strombus raninus*, and the Florida fighting conch, *Strombus alatus*. The first two are found in southeast Florida; the fighting conch is found from North Carolina to Florida. Conchs have a claw-like operculum on their narrow foot. Using the operculum as a vaulting pole, conchs push themselves forward in a series of small leaps.

Conchs are primarily herbivorous. With their large proboscis, conchs search out algae and edible debris in grass beds and reef areas. Adults have heavy shells with a thick expanded lip, but juveniles have thinner, less flaring shells. The wide lip helps shield the proboscis. Crustaceans, fish, and other snails prey on young conchs. Loggerhead turtles are known to crush adult Queen conchs, quite a feat considering the thickness of the conch shell. By far the most common predators of Queen conchs are people. Conch meat is used in salads, chowders, and fritters, and the shells end up in souvenir shops.

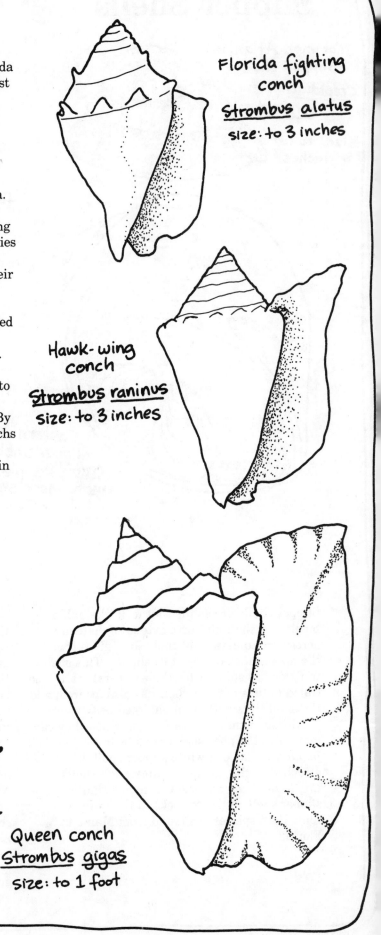

Florida fighting conch
Strombus alatus
size: to 3 inches

Hawk-wing conch
Strombus raninus
size: to 3 inches

Queen conch
Strombus gigas
size: to 1 foot

Moon Snails

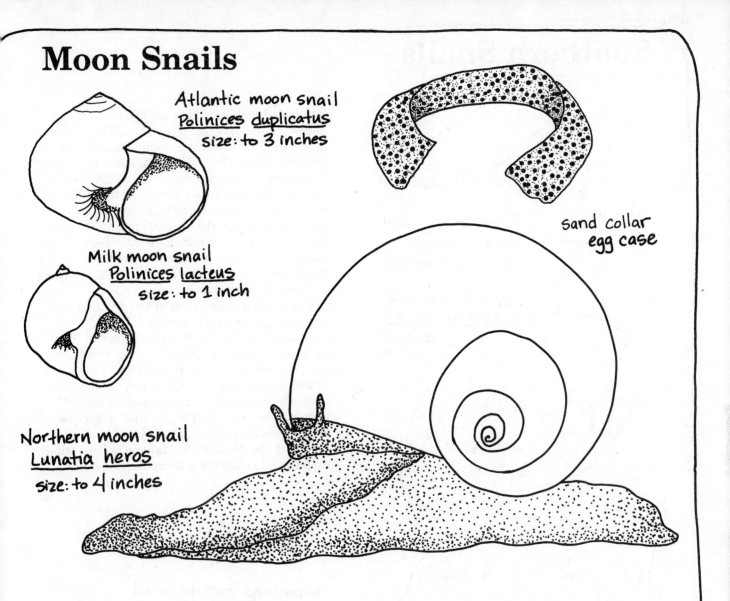

Atlantic moon snail
Polinices duplicatus
size: to 3 inches

Milk moon snail
Polinices lacteus
size: to 1 inch

Northern moon snail
Lunatia heros
size: to 4 inches

sand collar
egg case

Moon snails spend most of their time under the sand. With their large foot, these voracious carnivores plow beneath the surface of the sand in search of unsuspecting bivalves. Upon contacting a clam, the moon snail uses its strong foot to engulf the clam and hold it in place while it drills a hole through its shell. A gland at the tip of the snail's proboscis produces a substance that helps to soften the clam's shell. The moon snail then can drill through the clam shell with its specially adapted radula, leaving a neat round hole. Then it sucks out the soft tissue of the clam. Occasionally moon snails augment their diet of bivalves with various types of dead flesh. Some have a tendency to be cannibalistic.

Many species of moon snail live along the Atlantic Coast in lower intertidal and subtidal areas. *Lunatia heros,* the Northern moon snail, is found from Maine to North Carolina.

Although its shell is large, its slimy fleshy foot is larger. In fact, the foot looks like it could never fit back into the shell, but it does. The Atlantic moon snail, *Polinices duplicatus,* is found from North Carolina to Texas, while the Milk moon snail, *Polinices lacteus,* is found from North Carolina to Florida. All three of these snails have creamy colored shells. The Atlantic natica, *Natica canrena,* found from North Carolina to Texas, has a beautiful pearly shell with rows of light brown spots. Its foot is large but not as fleshy as the other species.

Eggs of moon snails are laid in gelatinous sheets that wrap around the parent shell, become covered with sand grains, and end up looking like a sand collar. Each species of moon snail has its own collar shape. Thousands of eggs may be in each collar. Embryos pass through their larval stages within the collar and hatch out as small snails.

Southern Snails

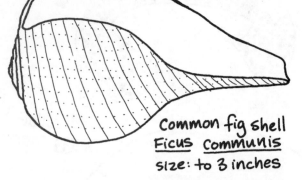

Common fig shell
Ficus communis
size: to 3 inches

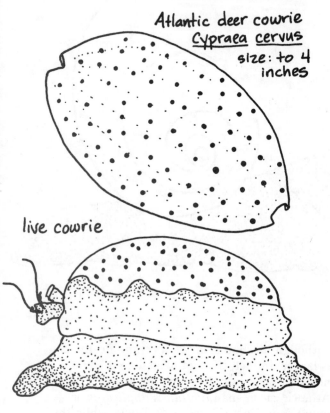

Atlantic deer cowrie
Cypraea cervus
size: to 4 inches

live cowrie

Illustrated here are three southern snails, none of which have an operculum. The Common fig shell, *Ficus communis,* found from North Carolina to the Gulf of Mexico, has a thin diaphanous shell that resembles a light bulb. Beneath the sculptured surface of the shell, the snail is visible even when completely withdrawn. When the fig shell crawls along in the sandy areas it inhabits, its pale fleshy mantle nearly covers the entire shell, much like the cowerie. Sea urchins and other echinoderms are their preferred food.

The smooth glossy shell of the Atlantic deer cowerie, *Cyprae cervis,* is often covered completely by the fleshy, bumpy mantle. The mantle helps protect the shell from scratches, scrapes, and general wear and tear. Young coweries have thin shells, but as they mature, their shells thicken and the outer lip curves inward. The last whorl of the shell overgrows all the previous whorls, obscuring the coiled nature of the adult shell. Many primitive tribes used cowerie shells as a form of money.

Many times the shell of a snail is more colorful than the snail itself. Not so with the Flamingo tongue snail, *Cyphoma gibbosum,* of the southeast United States. The color of the shell is not spectacular. In fact, the shell pales in comparison with the foot and the mantle, which are a light peach color with black-rimmed orange dots. As the snail crawls around, its beautiful mantle completely covers its shell. The shell looks like two thimbles glued together, open end to open end. Flamingo tongue snails live on sea fans and sea whips, eating live polyps.

Flamingo tongue snail
Cyphoma gibbosum
size: to 1 inch

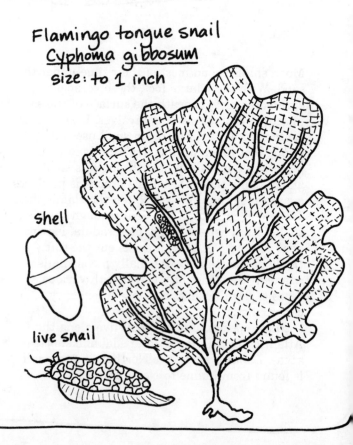

shell

live snail

Miscellaneous Snails

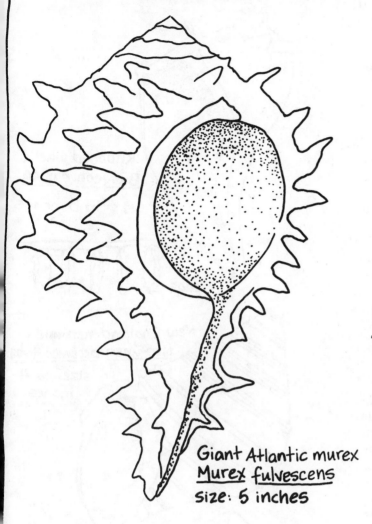

Giant Atlantic murex
Murex fulvescens
size: 5 inches

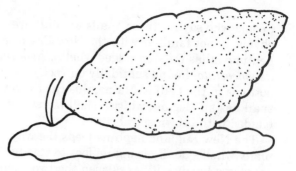

Blackberry drupe
Morula nodosa
size: to .5 inch

Atlantic oyster drill
Urosalpinx cinerea
size: to .7 inch

Many species of *Murex* live along the southeast Atlantic Coast. A gland in the foot of the murex produces a fluid that was used in ancient days as a purple dye. Murex shells often have beautiful spires and spines, as in the Giant Atlantic murex, *Murex fulvescens*. Found offshore from North Carolina to the Gulf of Mexico, this murex is carnivorous, as are all the murex snails. Using its foot to grasp a bivalve, it pries open the two shells with the outer lip of its own shell. Its radula then tears out the bivalve's flesh. Occasionally a murex may forsake bivalves in favor of newly dead creatures.

A relative of the murex is the Blackberry drupe, *Morula nodulosa*, a common intertidal snail found from South Carolina to Florida. Its textured shell is reminiscent of the beaded surface of blackberries and raspberries.

One of the biggest predators of oysters is the Atlantic oyster drill, *Urosalpinx cinera*. Almost no oyster from Nova Scotia to Florida is safe from this voracious carnivore. A secretion produced by a gland in the snail's foot helps soften the oyster's shell. The snail then drills with its radula. This process of applying the foot and then drilling with the radula is repeated until a hole reaches the soft tissue. This may take the better part of a day. Once the hole is completed, the oyster drill extends its proboscis and ingests the meat of the oyster.

Whelks

People who say northern snails are dull are obviously not familiar with the New England neptune (alias the Ten-ridged whelk), *Neptuna decemcostata*. Found subtidally on rocky bottoms in the Gulf of Maine, this carnivorous snail has an elegant, whitish shell tastefully banded with brown spiral cords.

The New England neptune feeds the way many whelks do. By using its foot to steady and envelop a bivalve, it forces open the two shells, using the edge of its own shell as a wedge. Then it inserts its proboscis and radula into the shell and consumes the contents. Lobstermen often find New England neptunes in their traps. These whelks scavenge as well as hunt, and are attracted to the dead fish used as lobster bait.

The Lightning whelk, *Busycon contrarium*, found from New Jersey to Florida, and the Knobbed whelk, *Busycon carica*, found from Cape Cod to Georgia, both feed in a manner similar to that of the New England neptune. They are not particularly voracious eaters, however; one large clam a month usually satisfies them. Females of both these whelks lay their eggs in long strings of plastic-like capsules. One end of the strand is attached to the bottom. Each capsule holds twenty to 100 eggs and has an escape hole near the top where the young snails exit upon hatching. The Lightning whelk often burrows in the sand, with its siphon sticking out of the sand to bring water over its gills.

A slightly less spectacular northern snail is the Atlantic dog whelk, *Thais lapillus*, found by the thousands on intertidal rocks from Maine to Long Island. This whelk is a carnivore. To get at the soft parts of its victims (especially Blue mussels) it drills through their shell with its radula, inserts its proboscis and ingests the body. Barnacles are another one of the dog whelk's favorite foods. The whelk forces open the top plates of the barnacle and eats the animal inside. The color of the dog whelk's shell is very dependent on the snail's diet. Dog whelks that eat many barnacles have white shells; those that eat more mussels have dark shells.

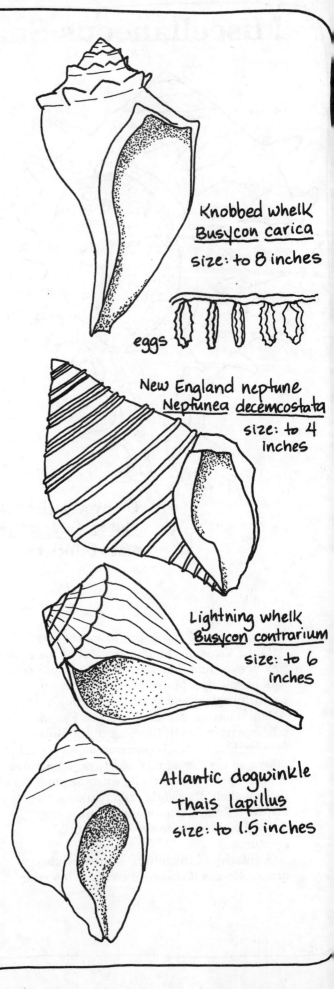

Knobbed whelk
Busycon carica
size: to 8 inches

eggs

New England neptune
Neptunea decemcostata
size: to 4 inches

Lightning whelk
Busycon contrarium
size: to 6 inches

Atlantic dogwinkle
Thais lapillus
size: to 1.5 inches

Cone Snails

Cones are tropical and subtropical snails with shells resembling ice cream cones. The Florida cone, *Conus floridanus,* is found from North Carolina to Florida, while the Alphabet cone, *Conus spurius,* is found only in Florida. Although these snails have brightly patterned shells, the colors are often hidden by the horny tissue layer, the periostracum, that covers the shell.

All cone snails have poisonous teeth on the radula which are used to capture prey. The teeth are calcified for rigidity, barbed at one end, and contain a toxin. Cones often lie partially buried in the sand. When prey wander by, the cone snail extends its proboscis and sticks the barbed end of one of the teeth into the victim. The teeth function singly, like darts or harpoons. The teeth may also be used for defensive purposes.

Most cone snails on the East coast eat worms. Once the worm has been impaled by a tooth, the injection of the toxin soon causes it to stop squirming. In this comatose state, the worm is consumed by the cone snail. A few cone snails along our coast feed on mollusks. As an unsuspecting snail crawls by a cone, the cone shoves a deadly poisoned tooth into the snail's foot. With the ensuing loss of muscle control, the snail becomes loosened from the columella of its shell. Then the cone holds its mouth over the opening of the snail's shell and ingests the snail as it goes limp.

Unlike the relatively innocuous cone shells of the Atlantic coast, certain species in the South Pacific are toxic and sometimes deadly to humans. Some Pacific species eat fish rather than worms and snails. As a fish pauses above a partially buried cone snail, the cone extends its proboscis up towards the fish and emerges from the sand to swallow its victim whole.

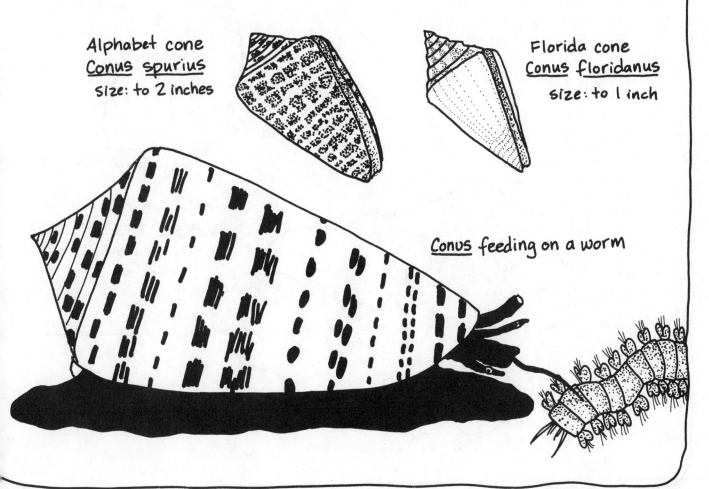

Alphabet cone
Conus spurius
size: to 2 inches

Florida cone
Conus floridanus
size: to 1 inch

Conus feeding on a worm

Bubble Shells

Bubble shells, sea hares and nudibranchs all belong to the subclass of gastropods known as the opistobranchs. Opistobranchs often have a reduced shell or no shell at all. Most have two pairs of tentacles. If they have gills, the gill is on the right side of the body.

Bubble shells are the largest group of opistrobranchs. Their shells are thin and fragile, like a bubble. While crawling along the bottom, the relatively large, yet delicate, bodies of the bubble shells nearly cover their shell.

The feeding habits of bubble shells are diverse. Although a few are herbivorous, most are carnivorous. They hook passing prey with their radula and swallow the victim whole. Bubble shells eat mostly snails and bivalves, which, once ingested, are crushed by calcareous plates in the bubble shell's gizzard.

The West Indian bubble, *Bulla occidentalis,* is white with brown markings and is found from North Carolina to Florida. *Haminoea solitaria,* the Eastern paper bubble, is amber colored and is found from Massachusetts to North Carolina. The Giant (relatively speaking) canoe bubble, *Scaphander punctostriatus,* is found along the entire coast.

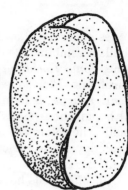

Eastern paper bubble
Haminoea solitaria
size: to .3 inch

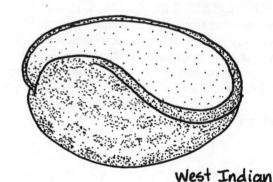

West Indian bubble
Bulla occidentalis
size: to .8 inch

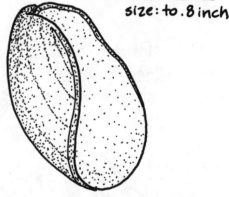

Giant canoe bubble
Scaphander punctostriatus
size: to 1 inch

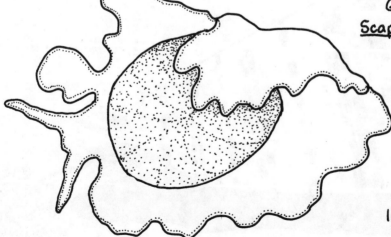

live bubble, crawling

Nudibranchs and Sea Hares

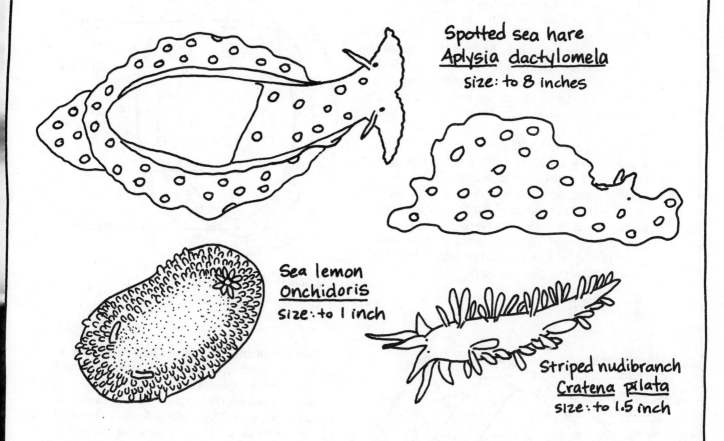

Spotted sea hare
Aplysia dactylomela
Size: to 8 inches

Sea lemon
Onchidoris
Size: to 1 inch

Striped nudibranch
Cratena pilata
size: to 1.5 inch

Nudibranchs, commonly called sea slugs, are gastropods without shells. They are found living intertidally or subtidally among sessile creatures such as tunicates, sponges, and hydroids. Unlike many other snails, nudibranchs have no proboscis, but they do have a radula, and many have jaws.

One group of nudibranchs, the eolids, have appendages called cerata all over their backs, which act as gills (nudi means "naked", branch means "gills"). The cerata increase the surface area of the skin, through which the nudibranch breathes. Each group of nudibranchs eat a particular type of prey. Eolids use their jaws to cut chunks of tissue from hydroids and anemones, but do not get stung in the process. The stinging cells of the prey pass intact through the nudibranch's digestive system to the tips of the cerata. Nudibranchs use the stinging cells to defend themselves. The Striped nudibranch, *Cratena pilata*, found along the entire coast, is an eolid nudibranch.

Another group of nudibranchs are the dorids. Dorids lack cerata but breathe with gill-like filaments that extend in a whorl around their anus. They specialize in eating sponges, bryozoans, and tunicates. A common dorid nudibranch all along the coast is *Onchidoris*, the Sea lemon.

Sea hares are very large gastropods with a thin shell buried under their skin, or no shell at all. They have a cauliflower-like gill under a flap of skin on the right side of their body. Two wide wing-like extensions of the foot stick out from their body. Some sea hares can swim by moving these flaps.

The Spotted sea hare, *Aplysia dactylomela,* is common in south Florida in spring when it comes into shallow water to lay its eggs, which may number over eighty million. Sea hares are herbivorous and use their jaws and radula to ingest algae. When disturbed, they squirt out a harmless, gooey purple ink.

Chitons

Chitons are armadillo-like mollusks that comprise the class Polyplacophora. Their flattened oval body is covered by a shell made of eight overlapping plates. Around the edge of the shell is the mantle, which in chitons is called the girdle. Both the foot and girdle clamp tightly to rocks so the chiton can withstand the turbulent waves and currents of the intertidal zone.

Chitons are notoriously lethargic mollusks. Usually they avoid the light and lay low during the day. At night they methodically crawl around scraping algae and minute organisms off rocks with their radula. Like limpets, chitons often return to their own resting spot when they are through feeding for the night. Chitons seem to stay motionless at low tide, waiting to forage when the tide comes in.

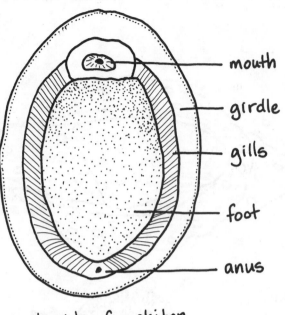

mouth

girdle

gills

foot

anus

underside of a chiton

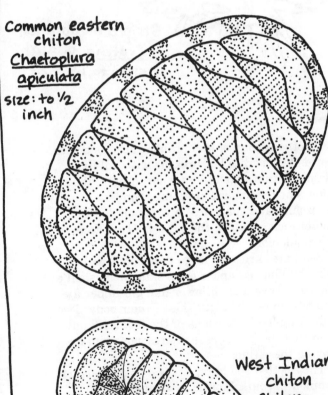

Common eastern chiton
Chaetoplura apiculata
size: to ½ inch

Because of their indistinct head and lack of tentacles, chitons are not particularly astute. However, some species have small, light-sensitive spots on their girdle that are sensitive to tactile stimuli. The main defensive strategy of the chiton is to adhere tenaciously to rocks, from which they are very difficult to dislodge. If removed, they roll up into a ball.

Approximately 600 living species of chitons are known today, but fossilized specimens dating back to the Paleozoic Era have been found. The Pacific coast has many more species and numbers of chitons then does the Atlantic coast. One giant Pacific chiton reaches a foot in length!

The Common eastern chiton, _Chaetopleura apiculata,_ is found from Massachusetts to Florida. On the New England coast, it is common in tidepools but in the south it is found subtidally.

One of the many chitons that lives intertidally from Florida to Texas is the West Indian Chiton, _Chiton tuberculatus._ Several species from the West Indies are used for human food and fish bait.

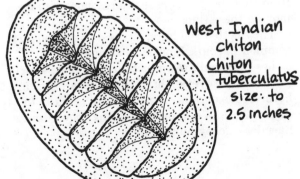

West Indian chiton
Chiton tuberculatus
size: to 2.5 inches

Bivalves

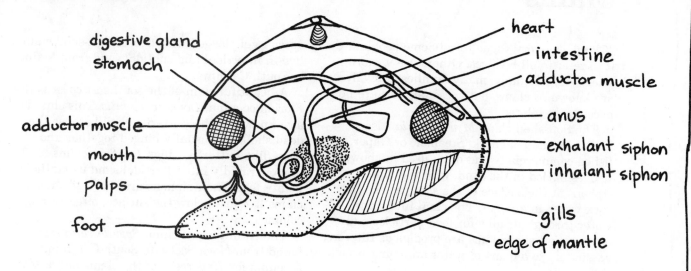

Bivalves are mollusks with a shell made of two halves, called valves, that are hinged together at the top. One or two large muscles keep the shell pulled shut. There are scars on the inside of the shell where these muscles are attached. When the muscles relax, an elastic ligament causes the valves to open. Along the hinge, interlocking teeth keep the valves from slipping. Between most of the shell and the mantle is a minute empty space into which new shell material is secreted. The mantle tissue attaches to the shell near the edge and leaves a scar there, the pallial line. The oldest part of the bivalve shell is the bump near the hinge, called the umbo.

The soft body of the bivalve is found within the shell. Bivalves have no head or radula. Usually they have a hatchet-shaped foot adapted for burrowing. As the bivalve extends its foot into the sand, the tip of the foot expands and acts like an anchor. Then the rest of the bivalve pulls itself down. Bivalves that do not burrow have a reduced foot.

In burrowing bivalves, the part of the mantle opposite the hinge is usually modified into two tubular siphons extending to the surface of the sand. One siphon brings water into the bivalve. Cilia on the gills create a current and send this water over the gills. There, mucus traps food particles, which are then sent in a groove to the mouth. Fleshy pads, the palps, near the mouth push the mucus-food mixture into the mouth. Naturally, the gills also extract oxygen from the water. Water is then expelled through the exhalant siphon. Most, but not all, bivalves are filter-feeders.

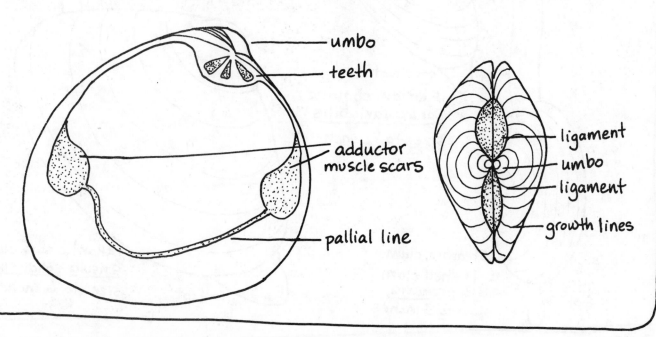

Clams

"Clam" is a very vague term. Some people use it to refer to all bivalves. Others use it more selectively. Actually, many families of bivalves are known as clams, but these families are not necessarily related.

The Soft-shell clam, *Mya arenaria*, also known as the Longneck, Steamer, or Gaper, is fished commercially and often used in chowders. It lives buried in the sand, with the tops of its siphons at the surface of the sand, where they suck water in and out. The clam filters the water through its gills for food particles and oxygen. Soft shell clams are prodigious filterers, passing up to a quart of water an hour through their body.

The long siphons of soft-shell clams are encased in a fleshy tube and cannot be pulled all the way back into the shell. The valves of the shell gape at both ends and cannot be completely closed. The mantle, however, is fused together around the outer edge, except at the siphons and foot, forming a continuous sheet of tissue around the shell opening so the clam doesn't get clogged with debris. Soft-shell clams live in intertidal and subtidal sand from Maine to South Carolina.

A beautiful clam of the southeast coast is the Florida coquina, *Donax variabilis*. This clam is also known as the Butterfly shell because the dead shells are usually hinged together and open, like butterfly wings. Coquina clams burrow rapidly into intertidal sand using their pointy foot. Their siphons are short, so these clams live just below the surface, often in huge colonies.

The Atlantic surf clam, *Spisula solidissima*, found from Nova Scotia to South Carolina, accounts for 70 percent of the clams harvested in the United States. Surf clams have heavy thick shells and short siphons. At low tide they are easily located by people and sea gulls. Surf clams live in intertidal and subtidal sand. Commercially, they are dredged from water thirty to eighty feet deep. Only the adductor muscle of this clam is edible.

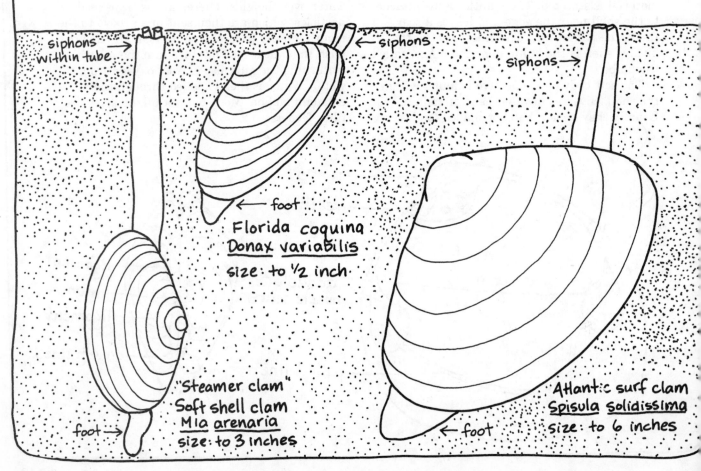

siphons within tube →

← siphons

siphons →

← foot

Florida coquina
Donax variabilis
size: to ½ inch

← foot

"Steamer clam"
Soft shell clam
Mia arenaria
size: to 3 inches

Atlantic surf clam
Spisula solidissima
size: to 6 inches

← foot

More Clams

New Englanders steadfastly refer to the Hard shell clam, *Mercenaria mercenaria,* as the Quahog (kō-hŭg). This succulent clam lives burrowed in intertidal and subtidal sand from the Gulf of Maine to the Gulf of Mexico. As with the Soft-shell clam, the whole Quahog, not just the adductor muscle, is edible. The Quahog has short siphons, consequently it burrows just below the surface, where it is easily collected with a rake or a shovel. Depending on the size of the clam, Quahogs are sold as Littlenecks (to one-and-a-half inches), Cherrystones (to two inches), and Chowder clams (to three inches or more). The inside of the Quahog is purplish and was used by the Indians for beads and currency.

One of the most agile clams is the Atlantic jackknife clam, *Ensis directus.* It is found from Canada to South Carolina, where it lives burrowed in intertidal mud flats. Because of its short siphons, the jackknife clam usually lives just below the surface. As the tide recedes, they

burrow deeper. Despite their unorthodox appearance, jackknife clams are edible, *if* you can catch them. They can burrow faster than you can dig. Jackknife clams can also swim, however erratically, by contracting their shell valves and stroking with their foot.

The Buttercup lucine, *Anodontia alba,* is a clam that lives in subtidal sand from North Carolina to Florida. Inside, the shell is a beautiful orange color. Outside it is white with light orange bands. Lucines live quite differently than most other clams. Using its long narrow foot, a Buttercup lucine burrows and makes a mucus-like sandy tube that serves as a tunnel for incoming water. Its foot may be six times the length of its shell, and periodically it pushes debris out of the tube. There is no inhalant siphon. Water exits through the exhalant siphon, which is very long, although able to be completely withdrawn into the shell.

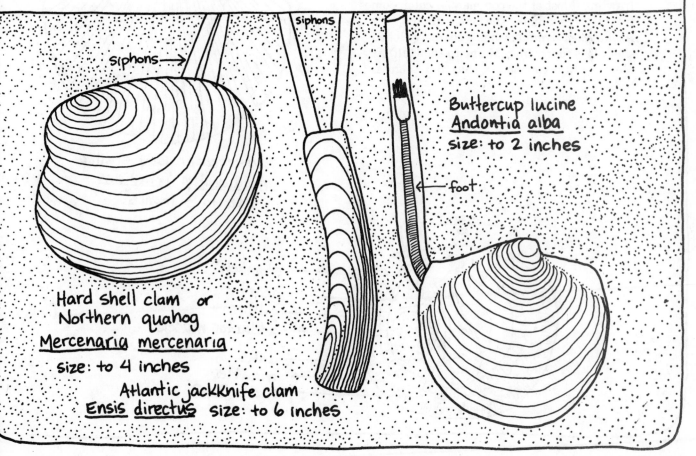

siphons

siphons →

siphons

Buttercup lucine
Andontia alba
size: to 2 inches

← foot

Hard shell clam or
Northern quahog
Mercenaria mercenaria
size: to 4 inches

Atlantic jackknife clam
Ensis directus size: to 6 inches

Scallops

Most bivalves live buried in the sand or attached to rocks; very few live freely on the bottom. One that does is the scallop. Scallops are mobile bivalves that live on the bottom, although not attached to it. They lie on their sides with their flatter valve face down. Since scallops do not burrow, they have no need for a powerful digging foot. Through evolution their foot has become greatly reduced.

Scallops have only one adductor muscle, but it is big and delicious. That muscle is the only part of the animal that people consume. When ordering scallops at restaurants, be aware that sometimes circular pieces of the fins of stingrays and skates are sold as scallops.

Scallops are fast swimmers With a quick clap of their valves, water is ejected from their mantle cavity and exits near the hinge, propelling the scallop forward. Swimming is used primarily to escape predators, especially sea stars. Scallops have beautiful blue eyes that line the frilled edge on their mantle. Although the eyes have a cornea, lens, and retina, they are not able to detect images, only shadows.

The Deep-sea scallop, *Placopecten magellanicus,* lives from Maine to North Carolina. Its shell is finely ribbed. Deep-sea scallops are harvested by dredging and trawling. Most of the commercial catch comes from George's Bank. The Bay scallop, *Aequipecten irradians,* has a coarsely ribbed shell. It is found from Maine to the Gulf of Mexico and is common in bays and Eelgrass beds.

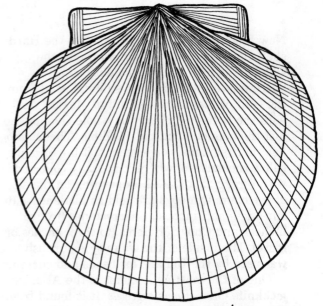

Atlantic deep-sea scallop
Placopecten magellanicus
size: to 8 inches

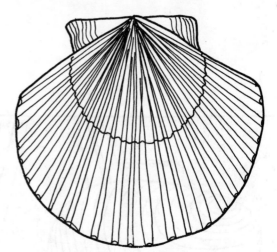

Blue-eyed or Bay scallop
Aequipecten irradians
size: to 3 inches

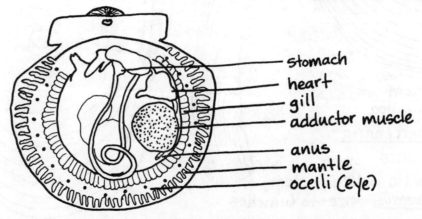

- stomach
- heart
- gill
- adductor muscle
- anus
- mantle
- ocelli (eye)

Mussells

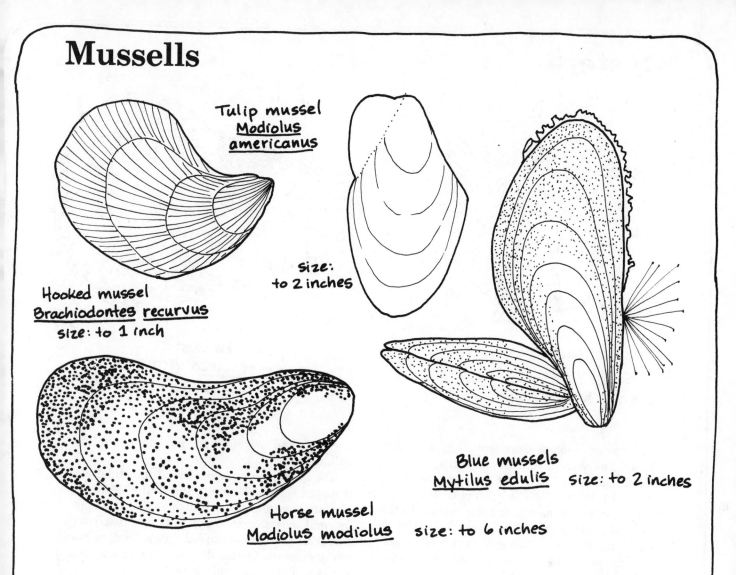

Tulip mussel
Modiolus americanus

Hooked mussel
Brachiodontes recurvus
size: to 1 inch

size: to 2 inches

Blue mussels
Mytilus edulis   size: to 2 inches

Horse mussel
Modiolus modiolus   size: to 6 inches

Mussels are another surface dwelling (non-burrowing) bivalve. They live attached to sea walls, rocks, and pilings by tough threads (byssal threads) secreted by a gland near their foot. The foot temporarily holds the mussel to the substrate while a secretion from this gland flows out along a groove in the foot. Gradually the fiber threads harden and the foot is withdrawn back into the shell. The threads are extremely tough, but not necessarily permanent. By letting go of the threads, mussels can move around on their small foot. Although their mantle is not modified into long siphons, mussels are filter feeders.

The Blue mussel, *Mytilus edulis,* grows in clumps and extensive beds in the intertidal zone from Maine to South Carolina. With its byssal threads, these mussels attach themselves to almost anything: shell fragments, pebbles, or rocks. Extensive colonies can even form in mud, as long as there are small particles beneath the surface to which the byssal threads can attach themselves. Life in the intertidal zone is tough, but Blue mussels are tougher. They can withstand freezing, excessive heat, and desiccation. When the tide leaves them high and dry, they breathe by passing air over their moist gills.

Blue mussels are edible, but are appreciated more in Europe than the United States.

A very common mussel offshore in New England is the Horse mussel, *Modiolus modiolus.* A shaggy black layer (the periostracum) covers its shell. The Horse mussel is found most often as a dead empty shell, often washed ashore in a kelp holdfast. Horse mussels are not edible.

The Tulip mussel, *Modiolus americanus,* ranges from North Carolina to Florida. It also has a shaggy periostracum covering its shell. The Hooked mussel, *Brachiodontes recurvus,* is common on pilings from Massachusetts to Florida.

Oysters

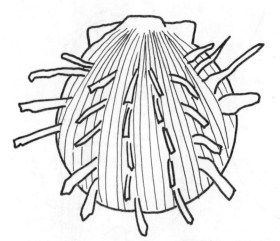

Atlantic thorny oyster
Spondylus americanus
size: to 5 inches

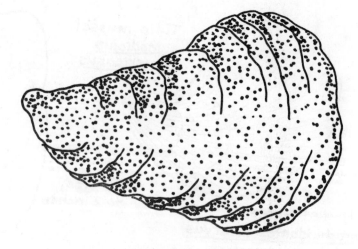

Eastern oyster
Crassostrea virginica
size: to 10 inches

True oysters such as *Crassostrea virginica*, the American oyster, live with one valve permanently cemented to a rock or other hard object. In the process of secreting their shell, young oyster larvae attach their mantle to the substrate with a drop of adhesive fluid. The oyster stays fixed to that spot for life. Oysters have no foot and no siphons. They open their shell to bring water over their gills.

All mollusks with shells can produce pearls. Pearls are formed as a result of an irritating particle becoming stuck between the shell and the mantle. The mollusk coats the particle with shell layers, and usually the particle becomes embedded in the shell. Particles that get covered with shell layers but do not become embedded become pearls. Pearls of edible oysters such as *Crassostrea virginica* are not nacreous and therefore are not commercially valuable.

The American oyster is found intertidally from the Gulf of Maine to the Gulf of Mexico. Oysters can tolerate salinities as low as 12 ppt (normal seawater is 35 ppt salt), and flourish in estuaries such as Chesapeake Bay. Oysters have many enemies: Oyster drill snails and sea stars are two of the most voracious.

The Thorny oyster, *Spondylus americanus*, is not a true oyster; it is a relative of the scallop. Like the oyster, it lives attached to the substrate by one of its valves. The Thorny oyster usually lives on coral or rocks. Like scallops, it has eyes at the edge of its mantle. The Thorny oyster lives in fifty to 100 feet of water from North Carolina to Florida.

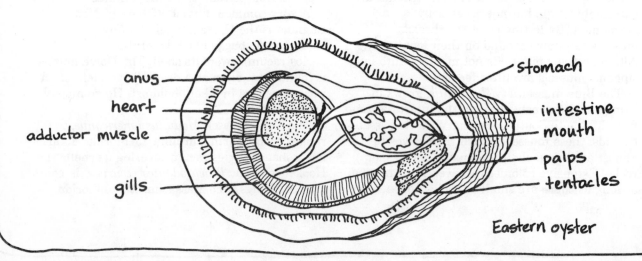

anus
heart
adductor muscle
gills
stomach
intestine
mouth
palps
tentacles

Eastern oyster

Ark and Jingle Shells

Arks are clams with strong box-like shells. Their shells are usually ribbed, dull-colored, and covered with a black periostracum. Unlike most other clams, not all arks live buried in the sand.

The Turkey wing or Zebra ark, *Arca zebra*, which is found from North Carolina to the Caribbean, lives attached to rocks below the low tide line. It attaches itself with a short, thick clump of byssal threads. The byssal clump is so large that there is a gap between the valves (opposite the hinge) for it. All arks have reduced siphons used in filter feeding. Many, such as the Turkey wing, have eyespots along the edge of the mantle. Living Turkey wings are often covered with encrusting plants and animals, but clean dead shells frequently wash ashore.

The Blood ark, *Anadara ovalis,* burrows in sand from Massachusetts to Texas below the low tide line. Only young ones form a byssus; adults do not. The Blood ark was named for its unusual blood. Most bivalves have no special respiratory pigment (the compound that carries oxygen) and have colorless blood. But the Blood ark has hemoglobin; consequently its blood is red.

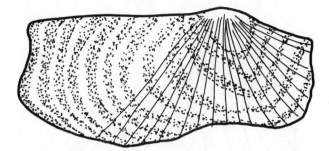

Turkey wing or Zebra ark
Arca zebra   size: to 3 inches

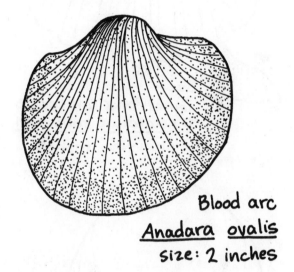

Blood arc
Anadara ovalis
size: 2 inches

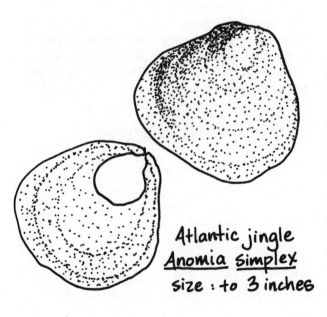

Atlantic jingle
Anomia simplex
size: to 3 inches

Jingle shells, *Anomia simplex,* are silvery translucent bivalves, sometimes known as Toe nail shells. They live attached to shells, rocks, and wood from the shore down to water thirty feet deep. Jingle shell valves are thin and flat, but grow to conform to the contours of the substrate. The bottom valve, which has a hole through it, is attached to the substrate on which the Jingle shell lives. Calcified byssal threads are secreted by a gland near the foot, and extend down through the hole and adhere to the rock. Jingle shells are probably most familiar as washed-up beach shells. Dead shells have been used in wind chimes and necklaces. *Anomia simplex* is found along the whole coast, along with several other species of jingle shell.

Cockles and Jewel Boxes

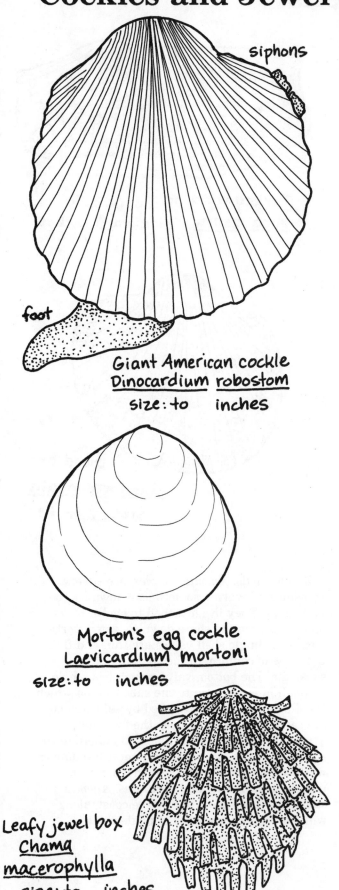

Giant American cockle
Dinocardium robostom
size: to   inches

siphons

foot

Morton's egg cockle
Laevicardium mortoni
size: to   inches

Leafy jewel box
Chama macerophylla
size: to   inches

Cockles are stately bivalves, often called heart clams because they are heart-shaped when viewed from the side. There are many species along the Atlantic Coast. *Dinocardium robostum,* the Great heart cockle, grows from North Carolina to Florida. Morton's egg cockle, *Laevicardium mortoni,* is found from Cape Cod to the Gulf of Mexico.

As do most clams, cockles live buried in the sand from the intertidal zone down to deep water. Their siphons are short, so they cannot burrow far beneath the surface. Their foot is sickle-shaped and muscular. Not only is it used in burrowing, but also for defense. When a predator approaches, the cockle literally leaps away by folding its foot and then pushing up off the bottom. Sea stars are particularly fond of cockles. As with so many other creatures, cockles are edible but popular as food only in Western Europe.

The Great heart cockle has a beautifully ribbed shell that is pink inside and yellowish outside. It lives buried in the sand below the low tide line, but often washes up as a dead shell. Morton's egg cockle burrows intertidally and subtidally in sand and sandy mud. It has a smooth shell and is sought by hungry ducks.

Jewel boxes are primarily tropical and subtropical bivalves. The Leafy jewel box, *Chama macerophylla,* is found from North Carolina to Florida subtidally out to water depths of a hundred feet. It lives attached to rocks, shells, or coral by its bottom valve. The bottom valve is deep, forming a box or cup. The upper valve fits over the lower valve like a lid; hence the name jewel box. The shell is quite variable in color and has a layer of long spines projecting from it.

Shipworms

Wooden docks have a tenuous hold on life. Winter ice can shear off their pilings. Hurricanes can reduce them to rubble. Ships can ram into them. But perhaps their most unrelenting enemy is the shipworm, *Teredo,* or *Bankia.*

Shipworms are not worms. They are worm-like bivalves that bore into wood using the serrated edges of their small shells. At the front end, shipworms have a foot and a two-valved shell. At the back end are siphons and pallets. Between the front and the back is the long fleshy body of the shipworm, which can reach one foot in length.

A shipworm larva settles on any submerged wood in salt water: boats, pilings, docks, or floating timbers. Using a cutting motion with its shells, it bores a pinhole large enough to accommodate its body and then continues to bore, enlarging the tunnel as it grows. The siphons and pallets remain near the pinhole opening. One siphon pumps water into the shipworm; the other pumps out waste water.

The paddle-like pallets seal off the pinhole when the siphons are withdrawn. Shipworms never widen the original pinhole entrance and they remain in the tunnel for life. If a shipworm is removed from its tunnel, it cannot excavate a new one.

Shipworms are covered entirely by their mantle, which secretes a thin shelly substance that lines the tunnels. Tunnels follow the grain of the wood. Eventually the wood becomes so riddled with shipworm tunnels that it completely disintegrates. Apparently the shipworm feeds on sawdust as it bores. Additional nutrition may be provided by food particles entering with water in the siphon.

Shipworms are found all along the coast and are especially prevalent in warm water. They are difficult to discourage, but coating the wood with creosote or non-corrosive metal alleviates some of the problem. Another creature that bores into wood is an arthropod called a *gribble* (see page 133).

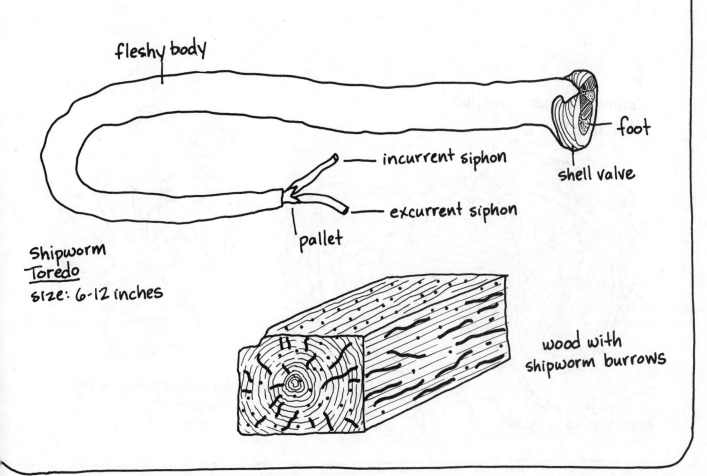

fleshy body

incurrent siphon

excurrent siphon

pallet

foot

shell valve

Shipworm
Toredo
size: 6-12 inches

wood with
shipworm burrows

Cephalapods

The most advanced and complex mollusks belong to the class Cephalopoda. Cephalopods include the Nautilus, the squids, and the octopi. The Nautilus of the southwest Pacific is the only cephalopod with a fully developed shell. Squid have a small internal shell. Octopi have no shell at all.

Unlike other mollusks, cephalopods are active swimmers. Water is drawn into their mantle cavity and then forced out through a siphon near the mouth. This propels the cephalopod in the opposite direction. The siphon can be pointed in any direction.

Cephalopods are predators. They locate prey with their eyes, which are very much like our own. Their eyes form images (not just shadows) and may also detect colors. Arms encircling the cephalopod's head are used to capture and hold prey. Once the prey is nabbed, the cephalopod bites into it with its beak. Then the radula brings the food into the mouth. Most cephalopods escape their predators by swimming, rapidly changing color, or squirting ink into the water through their siphon. The ink may confuse the predator, who probably has poor vision and can't discern a blob of ink from an octopus.

Most cephalopods are dioecious; males and females are separate individuals. Males have one arm modified to insert a sperm packet into the mantle cavity of the female.

The Argonaut or Paper nautilus, *Argonauta argo*, is a close relative of the octopus. Female argonauts have two specialized veil-like arms, which each secrete half of a beautiful parchment-like egg case. This shell is carried by the female, who uses it as a shelter for herself and as a brood chamber for her eggs. She is not attached to the shell, however, and may leave it at any time.

The male Argonaut is only half-an-inch long, but his sperm-carrying arm may be ten times his length. This long arm breaks off inside the female's mantle cavity, where it continues to wiggle around until the sperm are released. Early biologists thought this arm was a parasitic worm! Some male Argonauts live within the female's egg case.

Argonauts are pelagic and swim near the surface, where they prey on fish. The shell of the female washes ashore on beaches from New Jersey to Texas.

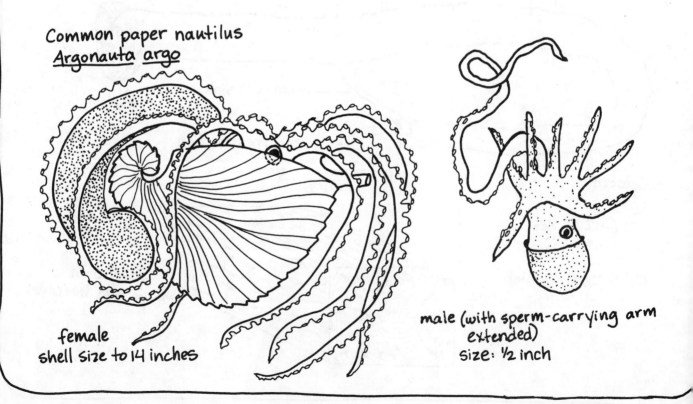

Common paper nautilus
<u>Argonauta argo</u>

female
shell size to 14 inches

male (with sperm-carrying arm extended)
size: ½ inch

Squids

Squid are the fastest swimmers among the invertebrates. Their bodies are sleek and aerodynamically shaped, allowing them to fly through the water (and occasionally through the air) with incredible agility and speed. Fins along the side of their bodies act as stabilizers. Out of the water, squid look soggy and flaccid—a sorry sight compared with their graceful shape under water. Squid found off the coast of the United States have a stiff internal shell called the pen.

Ten arms surround the squid's head. Actually, only eight of them are properly called arms. They are short and thick, with suction discs all along their inner surface. The other two arms are called the tentacles. They are thin and at least twice the length of the arms. The ends of the tentacles are spoon-shaped and covered with suckers.

Squid prey primarily on fish. As a school of fish swims by, the squid rushes in and grabs one with its tentacles. With its beak, it bites out a chunk behind the fish's head, or bites off the head entirely, and tears off large pieces of meat. The radula pulls the bites of food into its mouth. The squid usually leaves behind the fish tail and gut.

The Atlantic long-finned squid, *Loligo peali,* is abundant in shallow water from Canada to the Caribbean. It is fished commercially, and sold for food or bait. Eggs of this squid are laid in gelatinous strings on the bottom. Adults die after spawning. Another squid found along the entire coast is the Common short-finned squid, *Ilex illecebrosus.* It, too, is sold as fish bait, and is especially popular with cod fishermen. Both the long-finned and short-finned squid eat krill as well as fish. The Brief squid, *Lolliguncula brevis,* lives along the coast from New Jersey to Florida.

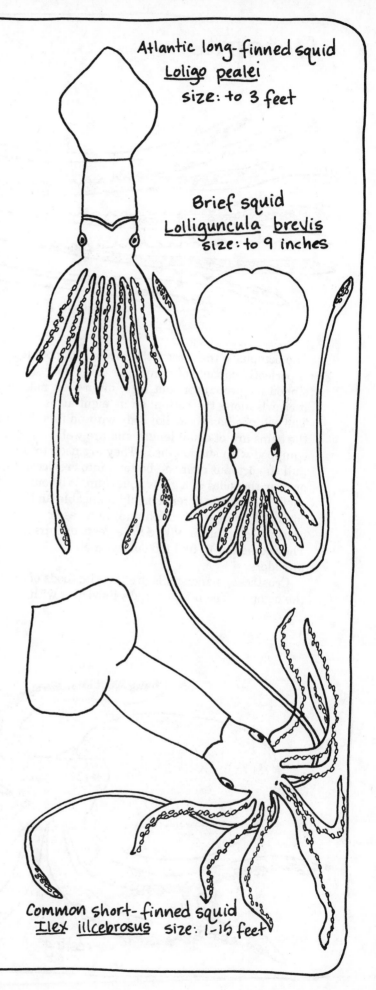

Atlantic long-finned squid
Loligo pealei
size: to 3 feet

Brief squid
Lolliguncula brevis
size: to 9 inches

Common short-finned squid
Ilex illcebrosus size: 1-15 feet

Octopus

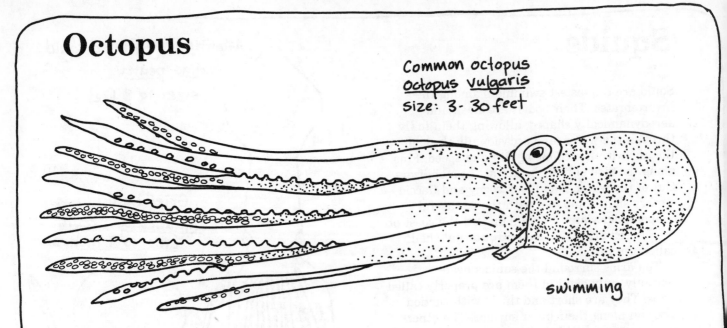

Common octopus
Octopus vulgaris
Size: 3- 30 feet

swimming

The octopus is the most lethargic cephalopod. It is perfectly capable of swimming using its siphon for propulsion, but more often than not, it crawls along the bottom on its eight arms. The octopus has a sac-like body with no fins. All the arms are of equal length and are well-equipped with suction discs. They are used to pull the octopus over the bottom, into crevices, or over the sides of a glass aquarium. Without a shell, an octopus is remarkably malleable and can easily squeeze through minute holes or cracks. Octopi usually live in retreats or lairs. They come out to find food or ward off intruders.

Crustaceans and snails are favorite foods of the octopus. The octopus grabs its victim with its arm and then injects it with poison from its jaws. Mucus from salivary glands and digestive enzymes begin to digest the prey even before it gets into the octopus' gut. A few species of octopus can drill holes through snail shells.

Male octopi have one arm modified to deposit sperm into the mantle cavity of the female. This arm has a scoop-like depression at the end. Females lay their eggs in clusters attached to caves, rocks, or empty shells. The female cares for the eggs until they hatch by keeping them free of debris and well-aerating them with streams of water from her siphon. Females of the Common octopus, *Octopus vulgaris,* die after the eggs hatch. The Common octopus lives from Connecticut to Florida.

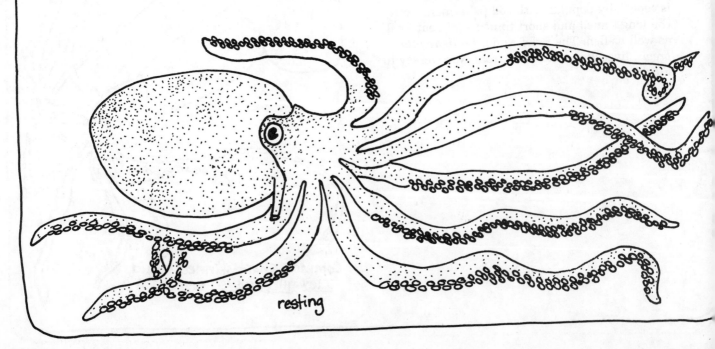

resting

Mollusk True/False Quiz

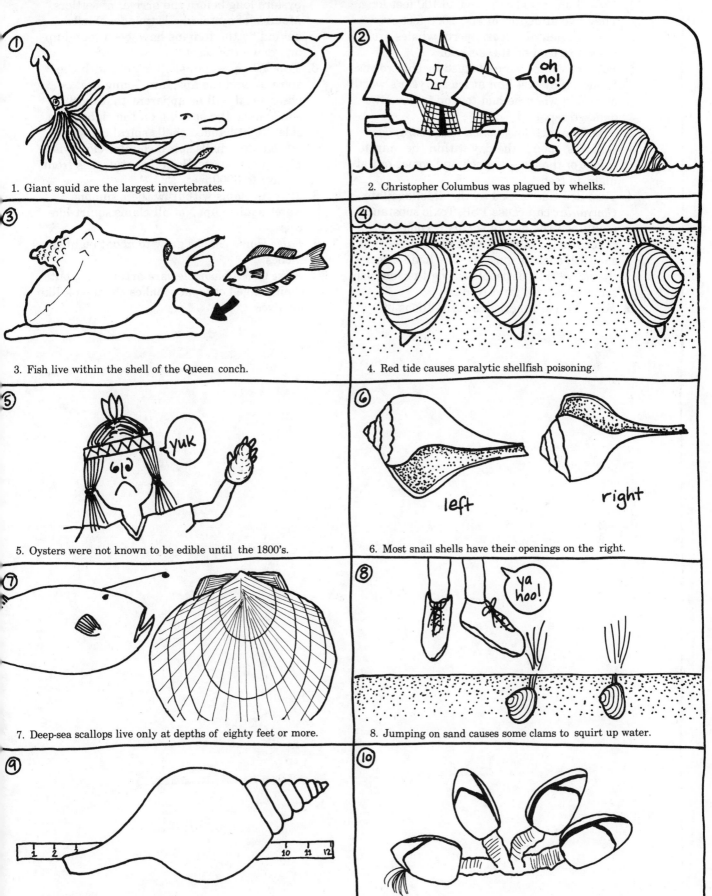

1. Giant squid are the largest invertebrates.

2. Christopher Columbus was plagued by whelks.

3. Fish live within the shell of the Queen conch.

4. Red tide causes paralytic shellfish poisoning.

5. Oysters were not known to be edible until the 1800's.

6. Most snail shells have their openings on the right.

7. Deep-sea scallops live only at depths of eighty feet or more.

8. Jumping on sand causes some clams to squirt up water.

9. Horse conchs can grow to three feet long.

10. Goose barnacles are mollusks.

1. *True* Giant squid can grow to 100 feet long. They live at depths of 900–2000 feet over the continental slopes. Sperm whales are known to feed on these squid.
2. *False* His ships were plagued by shipworms. In fact, they cost him at least a year's worth of sailing when he had to stop to have the damage repaired.
3. *True* The inch-long conch fish, *Astrapogon stellatus,* spends the day within the mantle cavity of the Queen conch and comes out at night to feed.
4. *True* Red tide caused by the dinoflagellate *Gonyaulax* can cause PSP. Toxic substances manufactured by *Gonyaulax* become concentrated in the tissues of filter-feeding shellfish such as mussels. The toxin affects the nerve cells of people eating the shellfish, sometimes with fatal results.
5. *False* Native Americans ate millions of oysters long before the arrival of settlers. Mounds of oyster shells (middens) left behind by the Indians have been found up and down the coast.
6. *True* When a snail shell is held with the spire up and the aperture facing the observer, it will be apparent that most snails have the opening on the right hand side. Right-handed shells spiral clockwise; left-handed shells spiral counter-clockwise.
7. *False* Deep-sea scallops live at depths from twenty to 3000 feet.
8. *True* As clams withdraw their siphons, water squirts up. Not all clams squirt like this.
9. *False* Horse conchs grow to about eighteen inches.
10. *False* Goose barnacles are arthropods, but their hinged carapace makes them look like bivalves.

Arthropods

Seventy-five percent of all animals belong to the phylum Arthropoda. Arthropods include the chelicerates (horseshoe crabs, spiders, and mites), the insects, and the crustaceans (crabs, shrimps, and lobsters). Nearly all marine arthropods are crustaceans.

All arthropods have an exoskeleton made of chitin, a protein. This external hard shell, in addition to being protective, gives rigid support for the attachment of the arthropod's muscles. The exoskeleton is made of separate plates connected by thin membranes. The segmented exoskeleton creates joints, allowing the arthropod to move its body and appendages. Due to this segmented nature of arthropods, they are believed to have evolved from an ancestral annelid (the phylum Annelida comprises the segmented worms). Although an arthropod grows, its exoskeleton does not, so the exoskeleton must be molted from time to time.

Arthropods have jointed appendages ("arthro" means joint, "pod" means leg) that perform a variety of functions, including swimming, walking, handling food, and gathering sensory information. Most arthropods have eyes, and most are dioecious (separate male and female individuals).

The largest group of marine arthropods is the subphylum Crustacea. Typically, the body of crustaceans is divided into two sections: the head and the trunk. The head has five pairs of appendages: two sets of antennae and three pairs of feeding appendages. The trunk is usually divided into the thorax and abdomen. Most often the gills are located where the legs are attached to the thorax. The exoskeleton of crustaceans is made of chitin *and* calcium.

Crustaceans usually brood their eggs. Upon hatching, the larvae look very different from the adults. With each successive molt, the immature crustacean develops a more adult-like appearance.

Lobsters, crabs, and shrimp, the most edible crustaceans, belong to the class Malacostraca. A typical malacostracan is illustrated below.

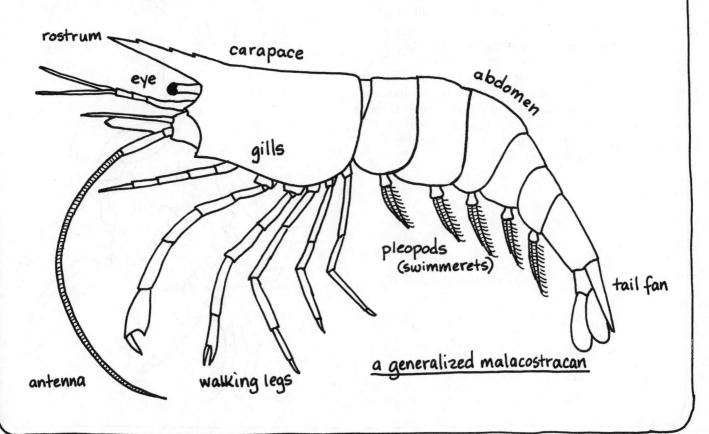

a generalized malacostracan

Molting

Since arthropods increase in size, while their exoskeleton does not, arthropods periodically molt their old exoskeleton to accommodate their expanding body. Molting is hormonally controlled.

The chitinous exoskeleton of arthropods is secreted by tissue layers underneath it. In preparation for molting, the tissue layer under the exoskeleton detaches from it and secretes a new exoskeleton. At this point the animal has two skeletons: an old outer one and a new inner one. When the new skeleton is completely formed, the old skeleton splits along specific weak points. For instance, the exoskeleton of Horseshoe crabs splits right along the front curved edge. Lobster exoskeletons split between the thorax and the tail. Crab exoskeletons split at the back. The animal pulls out of its old exoskeleton, leaving it entirely intact except for the split. Waste materials that accumulated in the old exoskeleton are also left behind. Arthropods then take in lots of water and/or air to stretch the pliable soft exoskeleton to a larger size before the tissues underneath harden it.

Crustacean molting has been especially well-studied. Immediately prior to a molt, crabs and lobsters become a bit sluggish. They usually seek shelter before molting, since their new soft skeleton will make them easy prey until it hardens. Mobility is limited immediately after a molt because the exoskeleton is not rigid enough to keep the limbs stiff. Many crustaceans eat their old exoskeleton to absorb the calcium salts in it. If a crustacean has lost a limb, the limb is gradually regenerated through successive molts. Although crustaceans molt throughout their entire life, they molt less frequently with age.

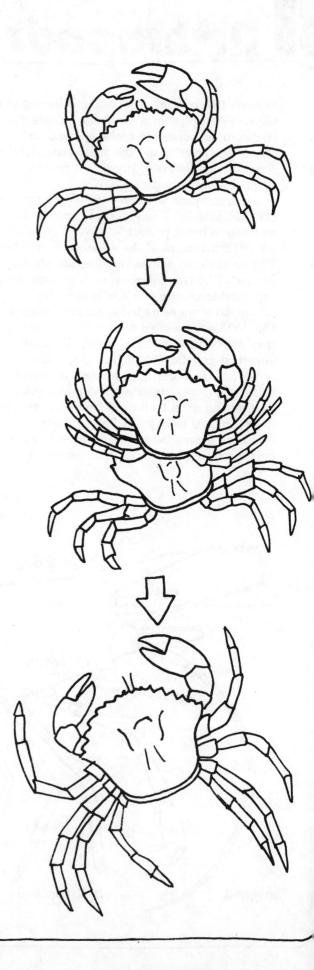

Sea Spiders

Few people are ever tempted to poke and probe among the plethora of hydroids, sponges and bryozoans that encrust every rock and piling at the edge of the ocean. But one good look at that tangled mass of marine life usually will reveal at least one sea spider—a slow, tiny, crawling arthropod. Sea spiders are common in all oceans, yet their small size (one-sixteenth to one-half of an inch), camouflaging coloration, and near motionlessness make them inconspicuous. Although anatomically different from true spiders, sea spiders superficially resemble them. They may or may not be derived from early marine arachnids that never left the sea; their evolution is unclear.

Sea spiders have three body segments: a head, a trunk, and a short abdomen. The head contains the proboscis, the mouth, two finger-like feeding appendages (chelicera) and four eyes. The trunk has long walking legs projecting from each segment. Male sea spiders have special legs used to carry eggs.

Many sea spiders have evolved with a prediliction towards eating the delectable hydroids, sponges, soft corals, and bryozoans on which they live. These carnivorous sea spiders may use their chelicerae to tear off pieces of their prey, or suck up tissues with their proboscis. Not all sea spiders eat meat; some feed on algae and detritus.

As the female sea spider lays eggs, the eggs are fertilized by the male, who in a rare display of paternal concern (relative to other marine creatures) gathers them onto his legs. There they are brooded and later hatch out as larvae. Gradually the larvae metamorphose into adults.

There are many sea spiders along the Atlantic coast. The Ringed sea spider, *Tanystylum orbiculare*, lives from Cape Cod to Florida. The Lentil sea spider, *Anoplodactylus lentus*, is found from Long Island to Florida. One species, the Anemone sea spider, *Pycnogonum littorale*, looks a little like a leech. It lives on sea anemones from Maine to New York.

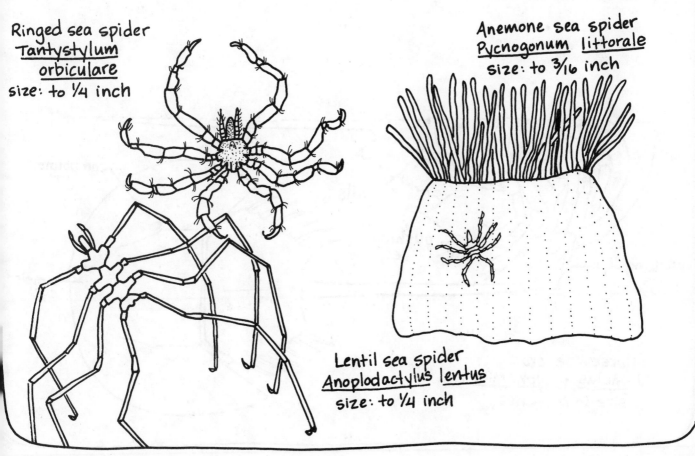

Ringed sea spider
Tanystylum
orbiculare
size: to 1/4 inch

Anemone sea spider
Pycnogonum littorale
size: to 3/16 inch

Lentil sea spider
Anoplodactylus lentus
size: to 1/4 inch

Horseshoe Crabs

Along with ticks, scorpions and sea spiders, Horseshoe crabs, *Limulus polyphemus,* are members of the subphylum Chelicerata. These formidable-looking yet totally innocuous creatures elicit shrieks of terror from beachgoers, who think they will be impaled by the long tail. Nothing could be further from the truth. As a Horseshoe crab plows slowly through the sand and muck, its tail acts as a rudder. And should the Horseshoe crab accidentally be turned over on its back, it sticks its tail into the ground and uses it to right itself.

On the top (dorsal) side of the Horseshoe crab, the two large compound eyes and two small simple eyes are easily visible. But don't miss the chance to turn over a Horseshoe crab to look at its underside. There you will see five pairs of legs. The first four pairs are used for walking. The last pair ends with flaps that open up and act like the disc on a ski pole, pushing the Horseshoe crab along without getting stuck in the sand. A long claw-like appendage, also on the end of the last pair of legs, is used to clean the gills, located behind the legs. The gills are folded like pages and are known as book gills. Horseshoe crabs can swim upside down, using their gills to propel them.

Horseshoe crabs feed primarily on burrowing mollusks and worms, but they are not averse to scavenging for detritus. The chelicerae pick up the food and place it in the ganthobase, where the legs attach to the center of the crab. The spiny first segments of the legs crush food and move it to the mouth.

To mate, the male Horseshoe crab, which is usually smaller than the female, hangs onto her back with special hooks on his first pair of legs. The female digs a hole in the sand and deposits the eggs, which the male fertilizes as they are deposited. When the larvae hatch out of the eggs, they are about half-an-inch long and similar to the adult, except their tail is very short.

Horseshoe crabs have countless commercial uses. At one time they were gathered and crushed for fertilizer and chicken feed, although the chickens' eggs tasted fishy afterwards. They are still used as lobster bait. An important medicinal use has been discovered for the Horseshoe crab's blue blood: Lysate, an extract from their blood, is used in cancer research and as an indicator of spinal meningitis.

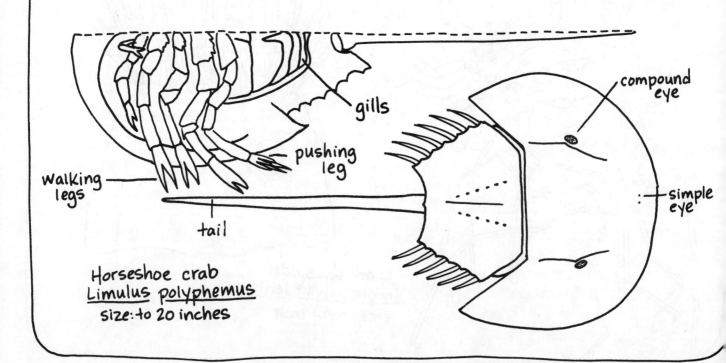

gills

pushing leg

walking legs

tail

compound eye

simple eye

Horseshoe crab
Limulus polyphemus
size: to 20 inches

Barnacles

Louis Agassiz, an American naturalist, once described the barnacle as a "shrimp-like animal standing on its head in a limestone house kicking food into its mouth." The most familiar types, the Acorn, *Balanus,* and Goose, *Lepas,* barnacles live on rocks, pilings, floating logs, and even on whales, shellfish, and penguin toes. Barnacles congregate quickly on almost any submerged surface in salty water, but cannot settle easily on fast-moving objects.

Barnacles are the only sessile (attached) crustaceans. Most are hermaphrodites—one barnacle has both male and female sex organs. But to propagate, most barnacle species must be fertilized by a neighbor. A retractable tube containing sperm reaches outside the shell as far as several inches to another barnacle.

Newborn barnacles emerge from their parents' shells as bristly, one-eyed larvae. After voraciously consuming plankton, they grow and molt into non-feeding, weak-swimming cyprid larvae, which soon settle to the bottom, creep around on their antennae, and search for a

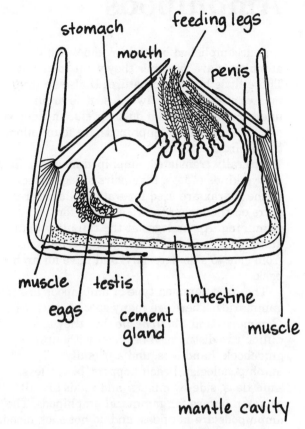

Acorn barnacle <u>Balanus</u>

Goose barnacles
<u>Lepas</u>

home. Within a few hours, a brown glue, now being studied by dentists for dental applications, anchors the larvae to the substrate. The larvae soon begin to metamorphose into adults.

Acorn barnacles secrete a volcano of overlapping calcareous plates, which totally encases them. Goose barnacles secrete a two-part shell and attach themselves to the substrate by a fleshy stalk. To feed, both barnacles use muscles to open their shell plates and extend their six feathery feet into the water, trapping plankton. The legs also have gills for gas exchange. The thin, soft cuticle covering the barnacle inside the plates is molted, but the hard plates themselves never are; they grow larger with the barnacle.

Amphipods

Any accomplished beachgoer knows what amphipods are, whether they want to or not. Those little bug-like creatures that leap from masses of decaying seaweed, scuttle around under rocks, and swim on their side in tidepools are all amphipods. Amphipods are found along the entire coast.

Literally translated, amphipod means *double*, or two kinds of legs. Five pairs of legs attached to the thorax are used for walking, and three pairs on the abdomen are adapted for swimming. Appendages at the tail allow some amphipods to jump. Amphipods are flattened from side to side (laterally) and have an arched back.

The most common type of amphipods are the gammarids. There are many species, such as *Gammarus* and *Talorchestia,* but they are difficult to distinguish between without a guidebook, hand lens, and a friendly amphipodologist. Beach hoppers, beachfleas, sand fleas, side-swimmers, and scuds are all common names for gammarid amphipods. The amphipods are not fleas, and do not suck blood.

Some gammarids burrow in tubes and are filter feeders. Others that are more vagarious scavenge for detritus. Gammarids that live in washed-up dead seaweed help to decompose it by eating the algae and the animals on it. The waste defecated by the amphipods is attacked by smaller animals and bacteria, which continue the decomposition process. The amphipods themselves are eaten in great quantity by fish, birds, and countless other creatures.

A less conspicuous and more bizarre type of amphipod are the caprellids, so-called "skeleton shrimp." They are very common on bushy seaweeds, bryozoans, and hydroids, where they move slowly and methodically, feeding on detritus, algae, or animals. Caprellids strongly resemble praying mantises in their form and behavior.

The last type of amphipods, the hyperiids, are planktonic and are often found living on jellyfish, which they may parasitize.

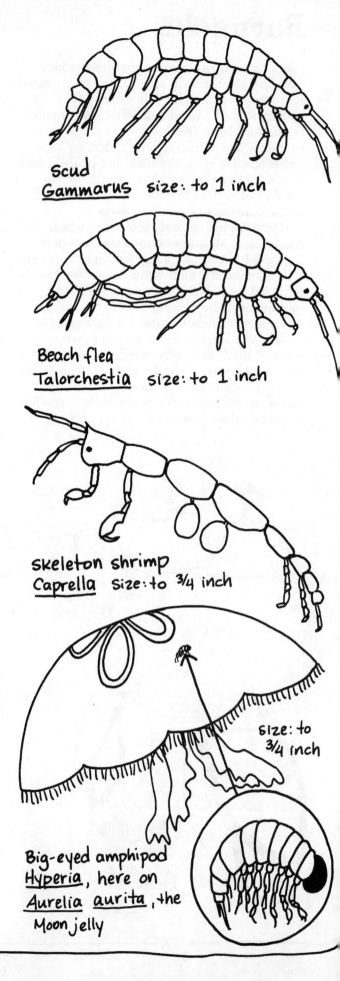

Scud
Gammarus   size: to 1 inch

Beach flea
Talorchestia   size: to 1 inch

skeleton shrimp
Caprella   size: to ¾ inch

size: to
¾ inch

Big-eyed amphipod
Hyperia, here on
Aurelia aurita, the
Moon jelly

Isopods

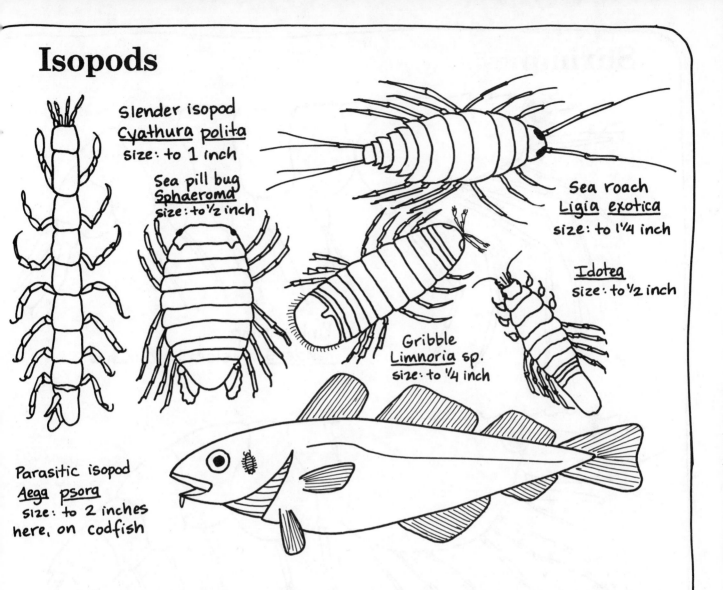

Slender isopod
Cyathura polita
size: to 1 inch

Sea pill bug
Sphaeroma
size: to ½ inch

Sea roach
Ligia exotica
size: to 1¼ inch

Idotea
size: to ½ inch

Gribble
Limnoria sp.
size: to ¼ inch

Parasitic isopod
Aega psora
size: to 2 inches
here, on codfish

Another small, widely-distributed arthropod is the isopod. Many marine isopods, such as the Sea pill bug, *Sphaeroma,* look a lot like their terrestrial cousins, the pill bugs and sow bugs. Isopods are flattened top to bottom (dorsoventrally) and have two compound eyes. Crawling legs are attached to their thorax, and pleopods on the abdomen act as gills. Isopods are usually a dull color.

Most isopods are bottom dwellers that can swim and crawl. *Idotea* is commonly found crawling and swimming among weeds, Eelgrass, and in tidepools. The Sea pill bug lives intertidally among hydroids, bryozoans, and seaweeds. A tropical species of *Sphaeroma* bores into the prop roots of mangroves. *Ligia* is the common sea roach of southern coasts. Hordes of frenzied sea roaches swarm over pilings and rocks in the upper intertidal zone. Many isopods such as *Cyathura polita,* the Slender isopod, burrow in sandy mud.

For the most part, isopods are scavengers or omnivores. *Limnoria,* known as gribble, eats fungus that grows in wood. Gribbles bore through pilings, docks, and driftwood in search of fungus, leaving the wood spongy and riddled with holes. Despite their small size, gribble can reduce a wooden dock to sawdust. As the damaged wood breaks off, gribble infest the wood underneath and are capable of doing as much damage as shipworms (see page 121).

A very specialized isopod, *Aega psora,* the Isopod fish louse (not a true louse), is parasitic in the skin, fins, or gills of fish, especially halibut and cod. They have huge eyes and piercing mouthparts, and may detach to swim free.

Shrimp

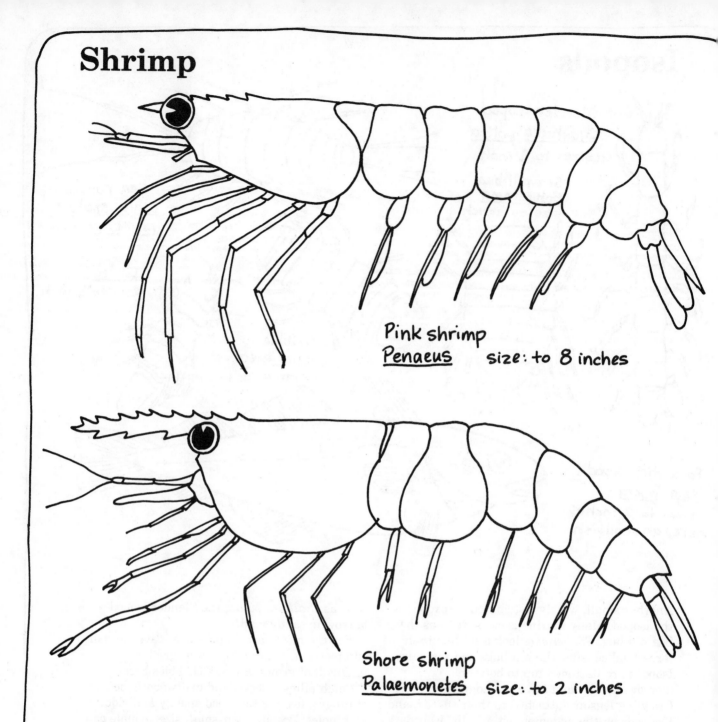

Pink shrimp
Penaeus   size: to 8 inches

Shore shrimp
Palaemonetes   size: to 2 inches

Shrimp, crabs, and lobsters all belong to the crustacean order Decapoda. "Deca" means ten and "poda" means leg; they have five pairs of legs. The first pair of legs usually has claws. Shrimp swim using their abdominal pleopods, although they are more often found crawling on the bottom than swimming about. Their feeding habits are eclectic. Some are carnivorous, while others eat plankton or detritus.

Compared to a crab or a lobster, most shrimp have a thin exoskeleton. Their gills, located where the legs are attached to the body, are hidden beneath the carapace. A special appendage near the shrimp's mouth beats to create a water current over the gills. Most

shrimp have a long spine, known as the rostrum, extending from the front of the carapace.

Penaeids, *Penaeus,* are the edible shrimp caught commercially off the southeast coast of the United States. No species are common north of Virginia, but a few hardy individuals stray to the gelid coast of Cape Cod. Penaeids burrow during the day, using their pleopods to excavate holes in the sand. At night they migrate to the surface. Shrimp, caught by trawling, are worth more per pound than any other seafood caught by U.S. fishermen.

Not-so-edible shrimp also abound. Shore shrimp are common on Eelgrass and on sand.

Shrimp

Snapping shrimp, *Alpheus,* are common south of Chesapeake Bay in shallow warm water and resemble miniature lobsters. The front margin of their carapace completely covers their eyes, although some light does get through. One of their claws may be as long as the snapping shrimp's entire body. When this oversized claw snaps shut, it creates a big banging noise, used primarily as a territorial threat to other snapping shrimp. The bang also stuns small edible fish and frightens predators. Snapping shrimp often live in burrows among shells and stones, or in the canals of sponges (see page 52).

Certain shrimp of Florida and the Caribbean are known as cleaner shrimp. The Red-banded coral shrimp, *Stenopus hispidus,* and Pederson's cleaning shrimp, *Periclimenes pedersoni,* use their claws to remove parasites from the scales of fishes. In this mutualistic relationship, the shrimp benefit by having a steady supply of food, and the fish benefit by being rid of annoying skin pests. Both the shrimp and the fish display certain behavioral signals (waving their antennae, remaining motionless, etc.) that indicate they are ready to clean or be cleaned.

Pederson's cleaning shrimp lives on sea anemones, while the Red-banded coral shrimp lives on coral.

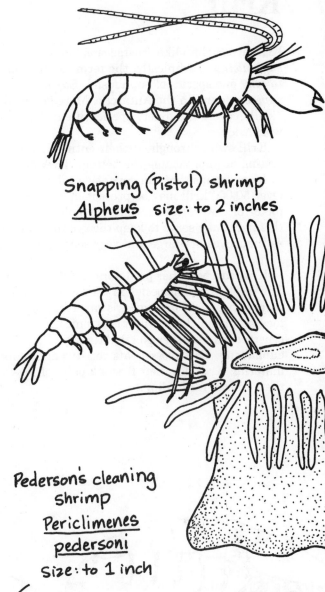

Snapping (Pistol) shrimp
Alpheus   size: to 2 inches

Pederson's cleaning shrimp
Periclimenes pedersoni
Size: to 1 inch

Red-banded coral shrimp
Stenopus hispidus
size: to 2 inches

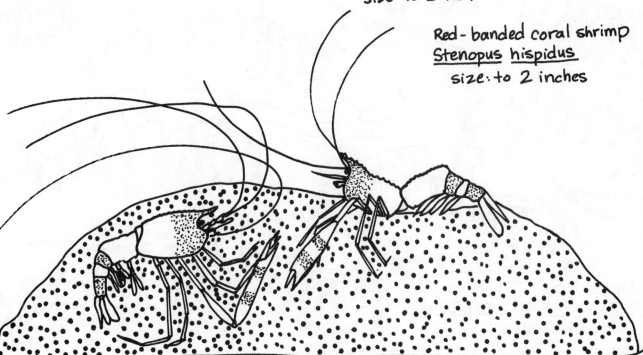

Krill

Krill are shrimp-like crustaceans known as *euphausiids*. Technically, the term "krill" refers to only one species of euphausiid, *Euphausia superba,* found in the Antarctic. More generally, "krill" covers the ninety species of euphausiids worldwide.

Krill swim throughout their entire life, never resting or approaching the bottom. They swim in huge groups that can cover several acres. Over 50,000 krill may be present in a cubic meter of water. Euphausiids are bioluminescent; this light may serve to keep the group together, as their eyes are large and very sensitive to light.

Krill can be either predators or filter feeders. The most common species in the Gulf of Maine, *Meganyctiphanes norvegica,* is carnivorous, feeding primarily on tiny planktonic crustaceans. Another North Atlantic euphausiid, *Thysanoesa,* eats detritus and plant material. Krill use their first six pairs of legs on their thorax for feeding.

Along the Atlantic Coast, krill are eaten by many creatures, including baleen whales, seals, pelagic fish, squid, and sea birds. Blue whales may eat four tons of krill a day. Euphausiids have a high protein content, up to 50 percent of their dry weight. For this reason, the Russians and Japanese have been harvesting krill for human consumption. They crush and press it, using its juice as a food enricher. The removal of all large baleen whales by hunting seems to have resulted in an increased krill population. But that doesn't mean krill can be harvested with reckless abandon. A trawler can net eight to twelve tons of krill per hour and could conceivably decimate a population. Krill is the only food for many animals, especially in the Antarctic.

Meganyctiphanes norvegica spawns in the summer and fall. The eggs are pelagic. Upon hatching, the nauplius larvae circulate with the water current. They molt many times and eventually metamorphose from a protozoea larvae to a zoea larvae to adult euphausiids.

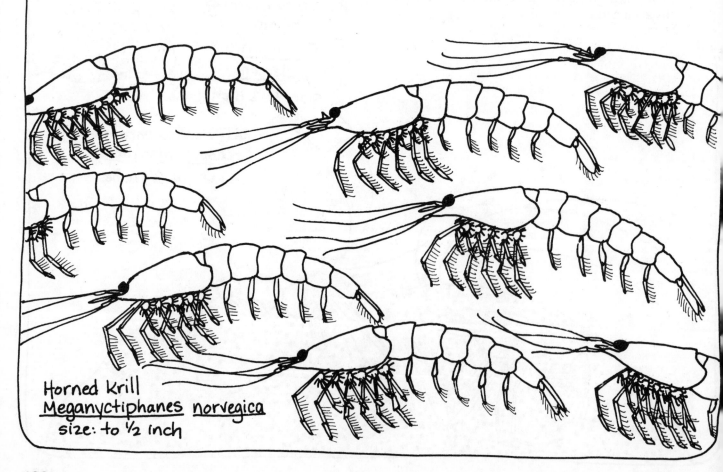

Horned Krill
Meganyctiphanes norvegica
size: to ½ inch

American Lobster

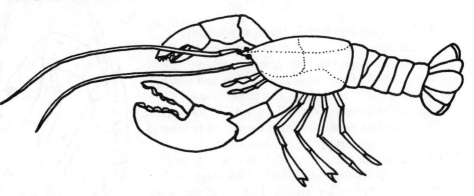

American lobster
Homarus americanus
size: to 4 feet
(usually much less)

Before the 1800's, American lobsters, *Homarus americanus,* were the most common large crustaceans off the New England coast. At low tide they could be caught with bare hands. Since the advent of commercial lobstering in the mid 1800's, however, American lobsters have become much more scarce. Each year, fishermen take nearly 90 percent of all legal-size inshore lobsters.

American lobsters are solitary, aggressive animals that live in rock ledges and crevices along the rocky bottom of the New England coast. Using their four sets of paired legs, lobsters walk along the bottom, feeling the path ahead by waving their antennae. They can swim by flipping their tail forward violently, which propels them backward. At night lobsters come out to scavenge for any type of organic matter from seaweed to dried fish. One claw is used to crush; the smaller claw is used to rip. Live food, such as clams and mussels, is also acceptable, but lobsters are rather lethargic and cannot catch fast-moving creatures. American lobsters are cannibalistic and will eat each other when confined to a trap for too long.

In mating, a male impregnates a newly-molted (still soft) female. Sperm is stored in the female's body until spawning, and remains viable for at least nine months. In spawning, eggs flow from an opening in the female over the sperm and are then attached to the mother's swimmerets (paddle-like appendages under the abdomen, the pleopods) by a natural adhesive. The eggs are aerated and protected under the tail throughout the nine–to–twelve month incubation period.

Most lobstering in New England takes place in the spring, summer, and fall. Wooden traps are usually homemade and the lobsterman burns his name into several laths. The hatch on top opens so the trap can be emptied and baited. There are two openings, each filled with a funnel-shaped cord net that allows the lobsters in but not out. Only legal-sized lobsters can be kept. The size varies slightly from state to state. Lobstering requires many special permits and licenses.

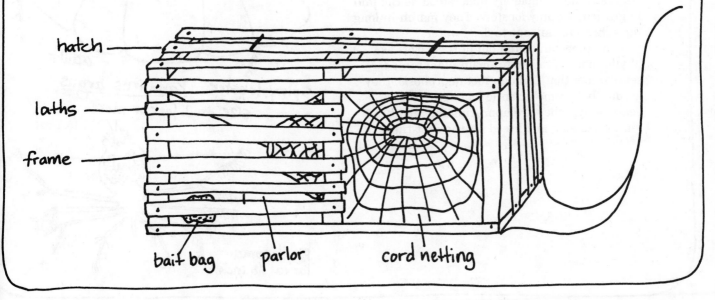

hatch

laths

frame

bait bag   parlor   cord netting

Spiny Lobster

New Englanders regard the Spiny lobster, *Panulirus argus,* with some skepticism, and rightly so. How can such a meek-looking creature, without massive crushing and tearing claws, be any relation to the pugnacious American lobster? Actually, the two "lobsters" are not that closely related; they even belong to two different groups of crustaceans.

Natives of the southeast U.S. coast, where the Spiny lobsters are found, often refer to the Spiny lobster as a crawfish. Instead of claws for defense, crawfish are covered with spines, especially on the carapace and antennae, and have two horns over their eyes. Trying to pick up a Spiny lobster is tricky. By thrusting their tail downward, they shoot backward, out of reach.

As with the American lobster, overfishing of the Spiny lobster has drastically reduced the population, and few large specimens are left. The most edible part of the Spiny lobster is its tail, which is usually what you get when buying lobster tails.

Spiny lobsters seem almost outgoing compared with American lobsters, whose longing for solitude drives them to cannibalism in close quarters. Spiny lobsters are quite gregarious. They congregate by day, taking shelter under rocks and ledges, with only their antennae protruding. But they will use their antennae to push away other lobsters if it gets too cozy. At night, they come out to find food, usually mollusks, worms, or dead animals.

Spiny lobsters have some unique traits. By rubbing the base of their antennae against a patch near their eye, they produce a loud, rasping sound that may scare predators. One trait, still not completely understood, is the fall migration of Spiny lobsters. They march in long lines, head to tail, out to deep water.

Young lobsters hatch out of their eggs as phyllosoma larvae and remain planktonic for several months. Through metamorphosis, they gradually become juvenile lobsters. At five years of age, they are approximately eight-to-ten inches long and breed for the first time.

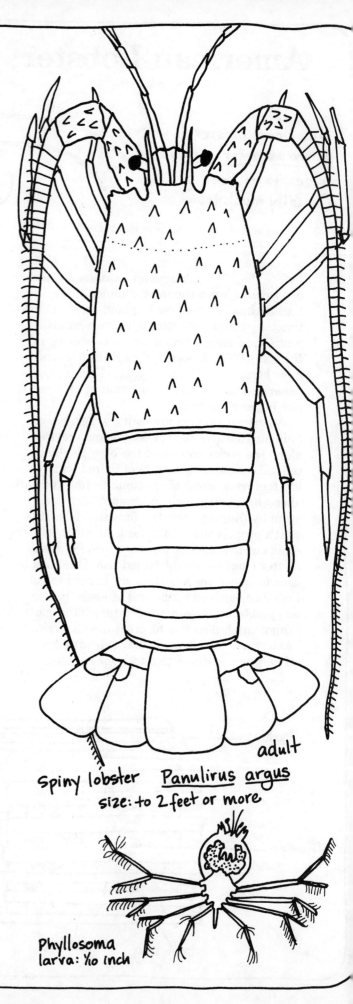

Spiny lobster **Panulirus argus**
size: to 2 feet or more
adult

Phyllosoma larva: 9/10 inch

Crabs

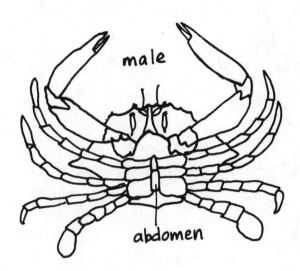

male

abdomen

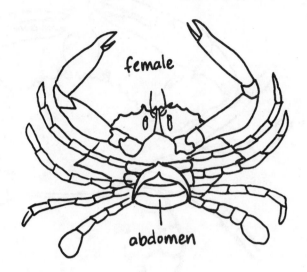

female

abdomen

True crabs are more or less folded-up lobsters. A crab's abdomen, which is very small compared with its body, is folded tightly beneath the body, where it remains permanently. By turning a crab over and examining its abdominal flap, it is usually easy to determine a crab's sex. Female crabs have a wide abdomen. Eggs are held there, between the body and the abdomen, during brooding. Males, with no need for a large abdominal flap since they do not brood eggs, have a thin, pencil-shaped flap.

All true crabs are in a subgroup of crustaceans called Brachyura. Horseshoe crabs and hermit crabs, along with some other "crabs," are not brachyurans and are not true crabs. In addition to the folded abdomen, true crabs have four pairs of walking legs and two legs with claws. They usually move sideways, holding their claws away from their body. Their bodies tend to be flat from top to bottom.

If you feel reticent about handling a live crab, pick up a crab's molt and take a close look at its features, such as the abdomen, claws, and legs. Notice the unique mouthparts. Two feeding appendages are modified to fit like doors over the mouth. True crabs have variable diets. They often scavenge for dead food, but if by a fortunate happenstance they catch a live animal, they will promptly devour it.

When male and female crabs mate, the female often stores the sperm until her eggs are ready to be released. When she releases the eggs, the stored sperm flows over them and fertilizes them. Then the eggs are attached in a big mass ("sponge") to the female's abdomen, where they are brooded until hatching. Eggs hatch into zoea larvae, and through successive molts gradually metamorphose into megalops larvae and then into adults.

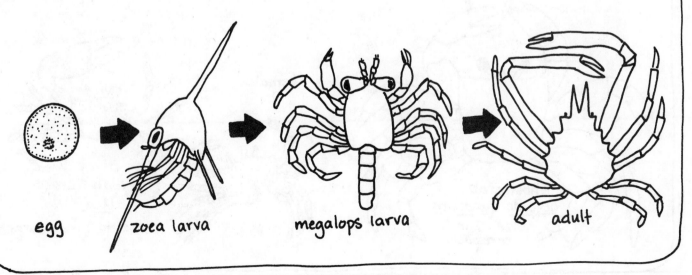

egg　　zoea larva　　megalops larva　　adult

Crabs

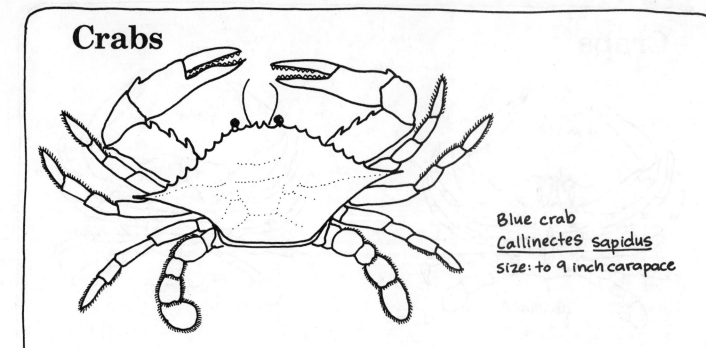

Blue crab
Callinectes sapidus
size: to 9 inch carapace

Crabs in the family Portunidae, such as the Blue crab, have paddles on the end of their last legs, enabling them to swim. *Callinectes sapidus,* the commercially important Blue crab, is found from Cape Cod to Florida, and is very common in estuaries. Blue crabs are fast swimmers and voracious carnivores, two traits that have led to the demise of countless small fish and helpless snails. Picking up a Blue crab involves more temerity than dexterity. These crabs can pinch!

A few crabs lead an amphibious existence. The fiddler crab, *Uca,* burrows intertidally in protected sand and mud beaches from Cape Cod to Texas. Male fiddler crabs have one claw greatly enlarged. The crab brings the excavated sand to the surface in small balls attached to its legs. At low tide, Fiddler crabs come out to eat detritus. As the tide comes in, they retreat to their burrows.

Fiddler crabs are gregarious and burrow very close to one another. This makes it easy for males and females to get together. Through an elaborate courtship ritual of claw-waving and rapping, males attract females to their burrows.

Another amphibious crab is the Ghost crab, *Ocypode quadrata,* found above the high tide line on sandy beaches from Delaware to Florida. Both the fiddler crab and Ghost crab have reduced gills capable of extracting oxygen from the air as long as they are kept wet.

Ghost crabs move with celerity, racing sideways, constantly changing directions. Their burrows are J-shaped and spaced far apart from each other. At night, Ghost crabs are very active, scavenging at the surf line for detritus or bivalves and wetting their gills at the same time.

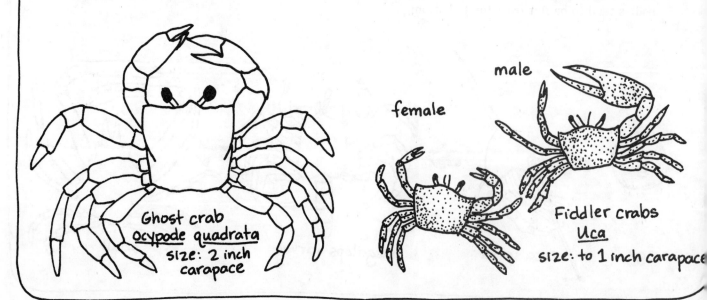

Ghost crab
Ocypode quadrata
size: 2 inch carapace

female

male

Fiddler crabs
Uca
size: to 1 inch carapace

Crabs

Two of the most common crabs found intertidally in New England are the Green crab, *Carcinus maenus,* and the Rock crab, *Cancer irroratus.* Green crabs live as far south as New Jersey and have a squarish, green carapace. The larger, but less common, Rock crab has a **D**-shaped, reddish carapace and is found as far south as South Carolina. South of Cape Cod, the Rock crab lives subtidally. The Green crab is the more corybantic of the two and is particularly active at night.

Both crabs forage in search of small worms, mollusks, and crustaceans, but are known to be scavengers as well as voracious predators. In turn, the crabs are eaten by gulls, herons, and fish. The Green crab is more tolerant of environmental extremes than the Rock crab. It can withstand brackish water, intertidal desiccation, and cold temperatures. Adults of both species, however, leave the intertidal zone in the winter.

The Stone crab, *Menippe mercenaria,* lives in rocky areas along the southeast and Gulf coasts. It is well known for its delicious meaty claws which can give a powerful pinch when still attached to the crab. This is one reason it is not wise to thrust fingers or toes into holes or under rocks. When fishing for Stone crabs, only one claw should be broken off and kept, because only the claws are edible. The crab is then released, and eventually it will regrow the lost claw.

Despite their gangly, spiny appearance, Spider crabs, *Libniia emarginata,* are harmless, slow-moving scavengers. They are found subtidally on all types of substrate along the entire coast. Being quite lethargic, Spider crabs prefer to defend themselves through camouflage rather than aggression. They attach algae and assorted debris to hook-like hairs on their carapace. Their claws end in cup-like depressions, appropriate for collecting the algae and detritus they consume.

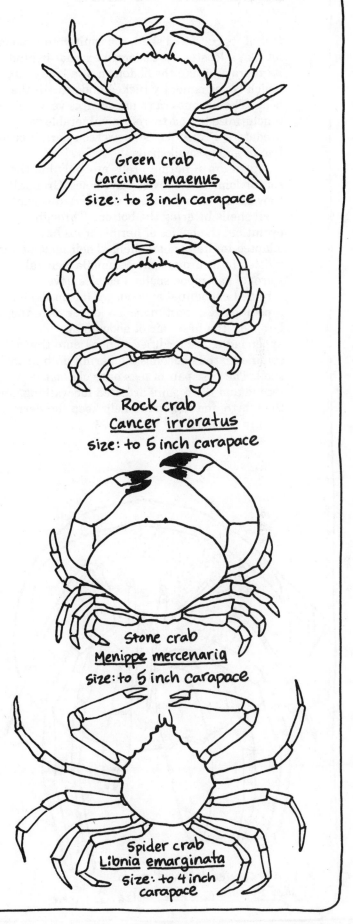

Green crab
Carcinus maenus
size: to 3 inch carapace

Rock crab
Cancer irroratus
size: to 5 inch carapace

Stone crab
Menippe mercenaria
size: to 5 inch carapace

Spider crab
Libnia emarginata
size: to 4 inch carapace

Hermit Crabs

Out of its borrowed shell, a hermit crab looks a lot like a lobster. Its abdomen extends behind its thorax, unlike the abdomen of a true crab, which is permanently tucked underneath the body. The hermit crab's abdomen is very soft (unchitinized), and therefore vulnerable to predators. Protection is one reason hermit crabs live in empty snail shells.

Ancestors of hermit crabs probably hid their soft abdomens in crevices and holes. Gradually they began to make use of the myriad vacant snail shells littering the bottom. Through evolution, the bodies of hermit crabs have adapted to life in snail shells. Their abdomen is twisted to fit around the columella (central spire) of gastropod shells. Females have retained abdominal appendages (pleopods) for brooding eggs; most males no longer have them. Uropods, the first pair of abdominal appendages, are modified to hook onto the columella, thus keeping the hermit crab in the shell. The first pair of legs have claws, the second and third pairs are used for walking, and the fourth and fifth pairs help keep the hermit crab in the shell.

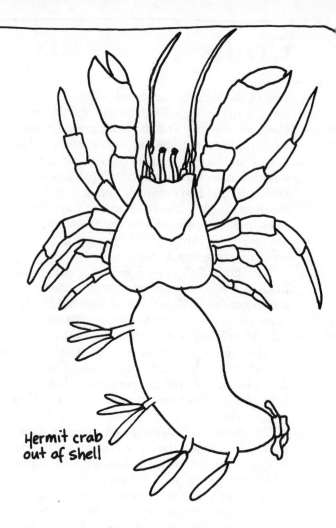

Hermit crab out of shell

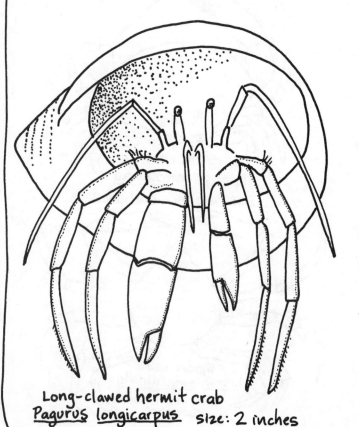

Long-clawed hermit crab
Pagurus longicarpus size: 2 inches

While scavenging for dead animals, the opportunistic hermit crabs are always on the lookout for a new shell, no matter how content they are with the one they have. In fact, the number of available shells limits the hermit crab population. Once hermit crabs spot an empty shell (rarely, if ever, will they kill a live snail), they check it out with their claws and antennae. If the shell seems appropriate they will get as close to the aperture as possible and instantly switch shells, always reserving the right to get out just as quickly as they got in. A hermit crab never leaves its shell before finding a new one. Of course, as the hermit crab grows, the shell does not, so even the less capricious hermit crabs must get new shells sometime.

The long-clawed hermit crab, *Pagurus longicarpus,* is a hermit crab found in shallow water all along the Atlantic coast.

Arthropod True/False Quiz

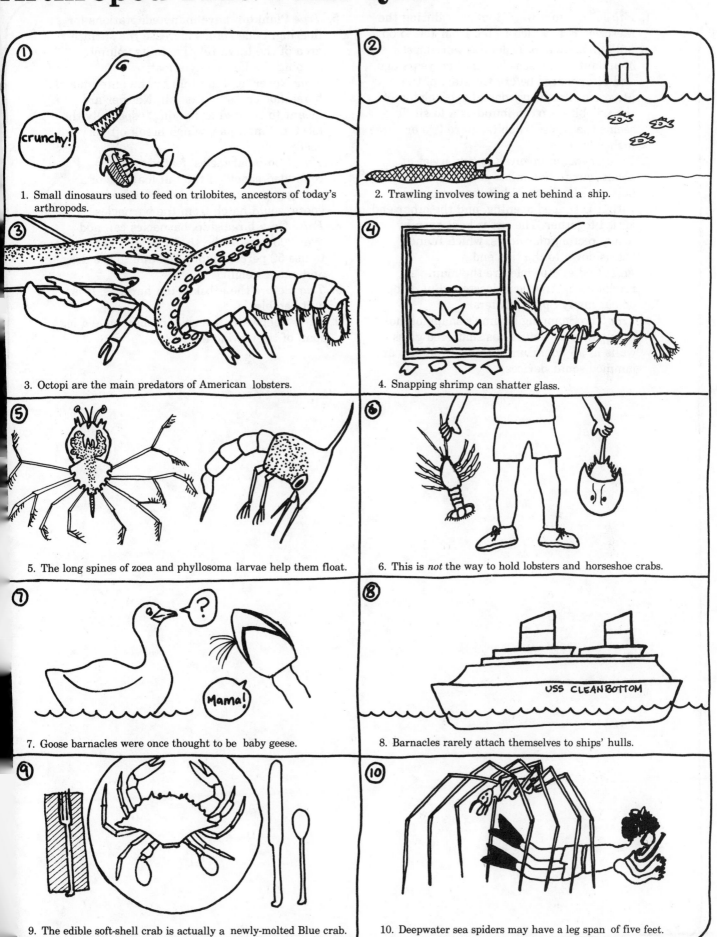

1. Small dinosaurs used to feed on trilobites, ancestors of today's arthropods.

2. Trawling involves towing a net behind a ship.

3. Octopi are the main predators of American lobsters.

4. Snapping shrimp can shatter glass.

5. The long spines of zoea and phyllosoma larvae help them float.

6. This is *not* the way to hold lobsters and horseshoe crabs.

7. Goose barnacles were once thought to be baby geese.

8. Barnacles rarely attach themselves to ships' hulls.

9. The edible soft-shell crab is actually a newly-molted Blue crab.

10. Deepwater sea spiders may have a leg span of five feet.

1. *False* Dinosaurs were dominant during the Mesozoic Era, which lasted from 190 to 65 million years ago. Trilobites were most abundant in the seas 550 million years ago and disappeared before the start of the Mesozoic. The heavy dorsal exoskeleton of the trilobite is often found as a fossil. It seems that most trilobites were two or three inches long.

2. *True* Trawling involves towing a net at about four to five miles per hour over the bottom. The front end of the net is wide, with a **U**-shaped opening, and the other end is tied together. The net is kept open by heavy rectangular doors, which funnel debris down to the tied end.

3. *False* Cod and people are the main predators of American lobsters. But octopi do eat many Spiny lobsters.

4. *True* The Snapping shrimp's snapping noise is so powerful that it has shattered glass walls in an aquarium, detonated mines, and jammed sonar devices.

5. *True* Plankton have many adaptations for floating. Long spines increase the surface area of the larva relative to its volume, helping the larva stay afloat.

6. *True* Never hold lobsters by the antennae or horseshoe crabs by the tail. Neither are meant to be used as handles, especially the lobsters' antennae, which break off very easily.

7. *True* Goose barnacles' feathery feet reminded people of the tail feathers of a young goose, and a rumor developed that a goose would hatch from the barnacles.

8. *False* On a large ship, barnacles can add over 200 tons to the ship's weight, causing it to use 50 percent more fuel due to the drag and extra weight.

9. *True* Before the exoskeleton hardens, it is quite edible.

10. *False* Deepwater sea spiders may have a leg span of twenty inches.

12 Echinoderms

Sea stars, brittle stars, sea urchins, and sea cucumbers are all echinoderms, familiar creatures to many beachgoers. The phylum Echinodermata dervies its name from the Greek words "echinos" and "derma," meaning "hedgehog skin." Most echinoderms have prickly, spiny, or warty skin. Ancestral echinoderms, as well as the larval stages of present day species, are bilaterally symmetrical: right and left halves are identical. Adult echinoderms, however, generally show a radial symmetry of five similar parts.

Unlike many other invertebrates, echinoderms are conspicuous, relatively large creatures. Most are benthic (bottom dwelling), and all are marine. There are no fresh water echinoderms.

Echinoderms have an internal skeleton covered with spines and skin. The skeleton is calcareous, but its form varies with the type of echinoderm. In sea stars and brittle stars, the skeleton consists of multitudes of small plates that move with one another, forming flexible joints. Sea urchins and sand dollars have skeletal plates that are fused, forming a rigid shell. In sea cucumbers, the calcareous plates have degenerated and are buried in the leathery, fleshy body.

Echinoderms possess a unique internal plumbing system known as the water vascular system. A hydraulic network of canals runs throughout the body, usually ending in a large series of tube feet on the surface of the animals. By varying the internal water pressure, the echinoderm can extend or contract the tube feet, which are used in locomotion, food collection, and respiration. Tube feet often end in small suction cups.

Echinoderms usually have their mouths underneath, on the oral side. The aboral side faces up. Most echinoderms have considerable ability to regenerate lost limbs.

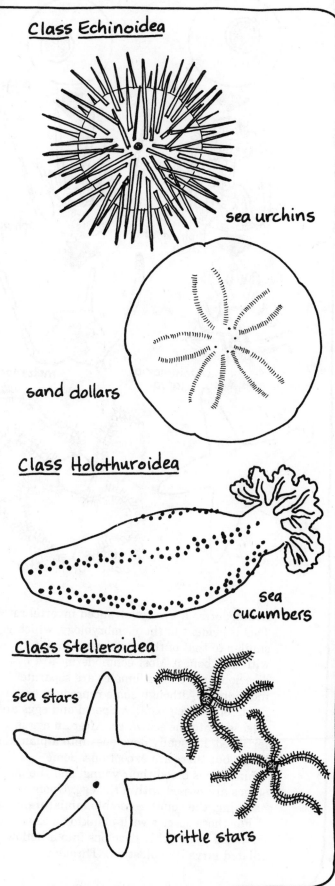

Class Echinoidea

sea urchins

sand dollars

Class Holothuroidea

sea cucumbers

Class Stelleroidea

sea stars

brittle stars

Life Cycle of Echinoderms

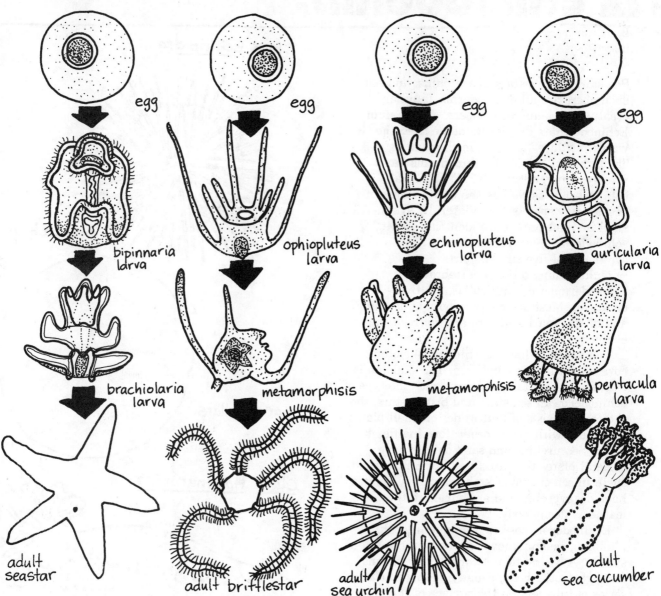

egg

egg

egg

egg

bipinnaria larva

ophiopluteus larva

echinopluteus larva

auricularia larva

brachiolaria larva

metamorphisis

metamorphisis

pentacula larva

adult seastar

adult brittlestar

adult sea urchin

adult sea cucumber

Echinoderms are very advanced invertebrates. This is evident in their embryology, which is similar to that of the vertebrates (animals with backbones). Most echinoderms are dioecious; males and females are separate individuals. Although some reproduce asexually, most reproduce sexually. Sperm and eggs are released into the seawater. After an egg is fertilized, it usually developes into a planktonic larva, but there are exceptions. Some echinoderms brood their young, and the larval stages are passed within the egg. Upon hatching, the young resemble miniature adults.

For those species with planktonic eggs, the fertilized egg quickly develops into a hollow ciliated larva (the blastula). Through metamorphosis, a pocket develops in the blastula larva, forming the gut. From this point on, the larvae of each echinoderm class look different (see above) but still have several things in common. All are bilaterally symmetrical and have bands of cilia used in swimming and feeding. Later, slender arms form, but they degenerate as the larvae gradually metamorphose into adults. Through a complex reorganization and degeneration of internal organs, the left side of the larva becomes the oral surface of the adult and the right side becomes the aboral surface. Only a small percentage of any invertebrate larvae survive to adulthood, since most are eaten, lost, or settle in the wrong environment.

Brittle Stars

Brittle stars are the most mobile, yet the most inconspicuous group of echinoderms. A few species burrow, but most shun daylight, taking refuge under rocks, seaweeds, and bottom debris. There they move around on their long snaky arms, scavenging for food.

Like their close relatives, the sea stars, the brittle stars have a skeleton of calcareous plates (ossicles) embedded beneath their skin. These ossicles are arranged in a line down the arm and each is covered by four small shields, giving the brittle star's arms an armored, jointed appearance. Spines and a pair of tube feet extend from tiny pores between the shields.

Brittle stars are rapid crawlers. They move by extending one or two arms forward, while the others trail behind and/or push against the bottom. The arms are very flexible and can coil over objects, but they are expendable. Should a predator lunge for a brittle star, the brittle star will cast off its arm, leaving the predator with a small snack as the echinoderm escapes. Arms can be regenerated as long as most of the central disc in intact.

Under the flat central disc lies the brittle star's mouth, framed by large-toothed plates that form a set of jaws. Most brittle stars are scavengers. They use their tube feet to collect debris from the bottom.

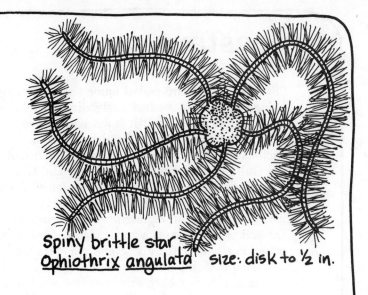

Spiny brittle star
Ophiothrix angulata   size: disk to ½ in.

The Spiny brittle star, *Ophiothrix angulata*, is found subtidally from Chesapeake Bay to Florida. The Short-armed brittle star, *Ophioderma brevispina*, lives in shallow water from Cape Cod to Florida. The most common intertidal brittle star north of Cape Cod is the Daisy brittle star, *Ophiopholis aculeata*. Basket stars have five branching, tangled arms that catch plankton drifting by. This basket star, *Gorgonocephalus arcticus*, is found north of Cape Cod, but there are subtropical species common on sea whips in Florida.

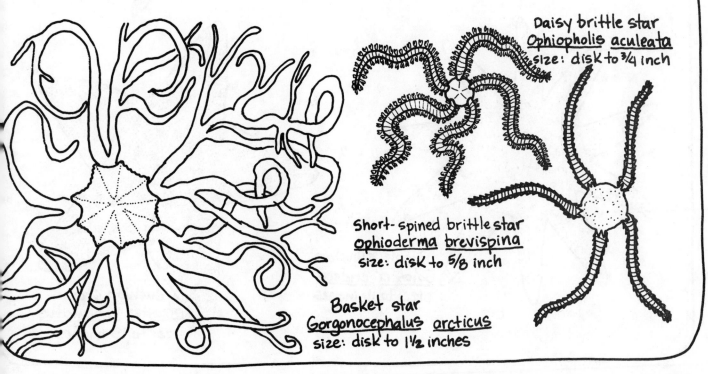

Daisy brittle star
Ophiopholis aculeata
size: disk to ¾ inch

Short-spined brittle star
ophioderma brevispina
size: disk to ⅝ inch

Basket star
Gorgonocephalus arcticus
size: disk to 1½ inches

Seastars

"Starfish" is an antiquated name that is quickly being replaced by the less misleading term "sea star." (After all, a starfish is not a fish.) Sea stars usually have five arms although some, such as the Purple sea star, *Solaster endeca,* of the Gulf of Maine, have nine or ten. The West Indian sea star, *Oreaster reticulatus,* found from Cape Hatteras to Florida has four to seven arms. Sea stars can regenerate their arms. At the end of each arm is a small pigment eye spot which is sensitive to light. Some sea stars are attracted to light; others avoid it.

The sea star's mouth and tube feet are found on its oral surface. Tube feet are clear and thread-like, and end in a sucker. Two or four rows of tube feet extend from ambulacral grooves in the middle of each arm. Water is taken into the water vascular system through the madreporite, a small, perforated scab-like plate on the aboral surface. That water is directed into canals in the arms and then into the tube feet. As water is forced into the tube feet, the tube feet extend. Sea stars move by adhering their tube feet to the substrate and pulling themselves along. Tube feet are also used in gas exchange.

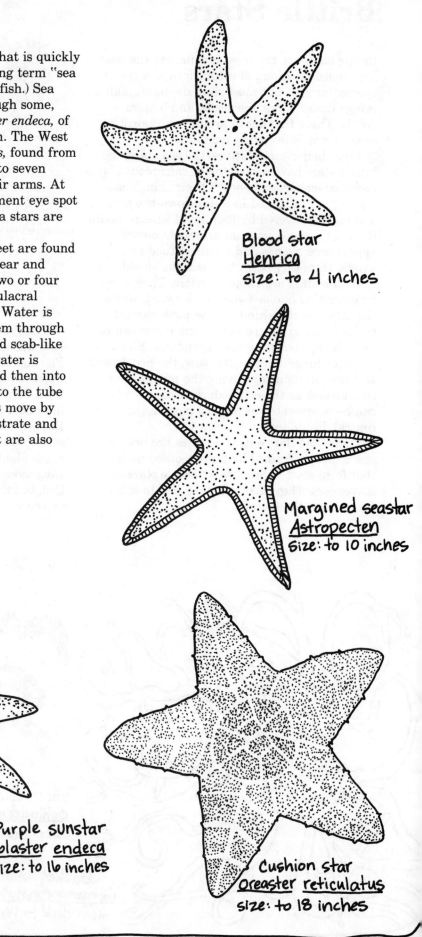

Blood star
Henrica
size: to 4 inches

Margined seastar
Astropecten
size: to 10 inches

Purple sunstar
Solaster endeca
size: to 16 inches

Cushion star
Oreaster reticulatus
size: to 18 inches

Seastars

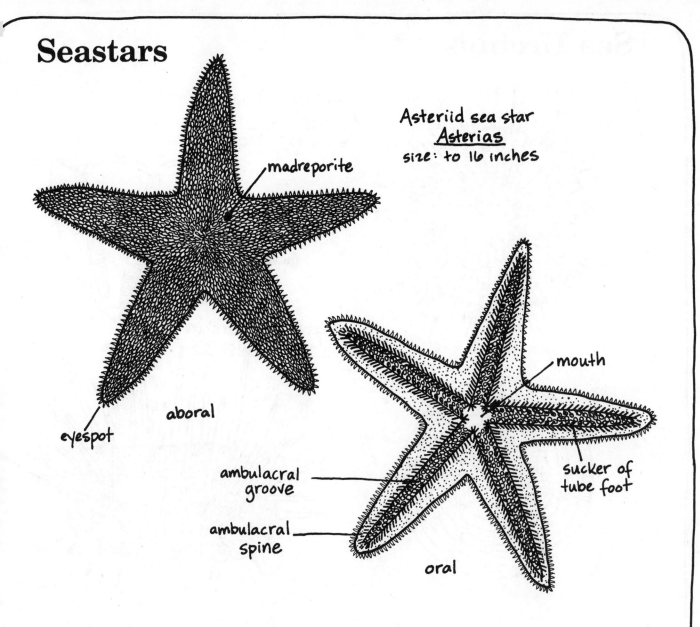

madreporite

Asteriid sea star
Asterias
size: to 16 inches

aboral

eyespot

mouth

ambulacral groove

sucker of tube foot

ambulacral spine

oral

As with brittle stars, the skeleton of a sea star is made of embedded ossicles that form an internal framework to support connective tissue. Bumps and spines of the ossicles give an uneven texture to the sea star's skin. The skin is also covered with mucus glands, sensory cells, and cilia. The mucus traps debris and dirt that fall on the sea star, and the cilia get rid of them. Many sea stars also have tiny pincers, known as pedecellarea, on their body surface. The pedecellarea can pick up and remove dirt and larvae that may try to settle there.

All sea stars are predators. *Astropecten,* the Margined sea star of the southeast coast, has tube feet without suckers. With its tube feet, it pushes through the sand digging for buried prey. *Henrica,* the blood star found from Maine to Cape Hatteras, is a suspension feeder. Plankton and detritus that float over its body are trapped in mucus. Cilia then sweep the particles into the ambulacral grooves and into the mouth.

Many sea stars eat with their stomach outside their body. Species of *Asterias,* the most familiar sea star along the coast, prey on worms, crustaceans, and bivalves, especially oysters. With their tube feet, they pull apart the two shells of the bivalve. A very narrow slit between the shells in all that is necessary. The sea star pops out its stomach through its mouth, keeping it attached to its body, and pops it into the bivalve shell. Digestive juices liquidate the victim, and cilia transport the soupy bivalve into the sea star's body.

Sea Urchins

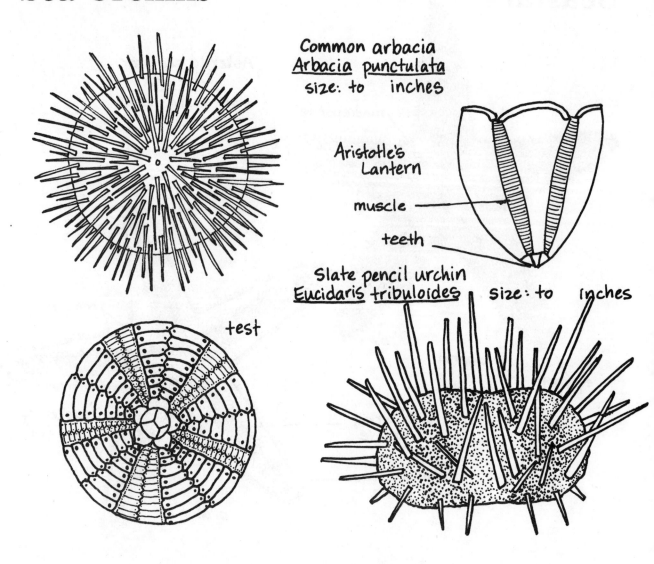

Common arbacia
Arbacia punctulata
size: to   inches

Aristotle's
Lantern

muscle

teeth

Slate pencil urchin
Eucidaris tribuloides   size: to   inches

test

Despite their spines, most sea urchins can be picked up and held, if it's done carefully. A notable exception is the Long-spined sea urchin, *Diadema antillarum,* of south Florida and the Caribbean. Its spines easily penetrate human skin, where they become embedded and break off. But even the Long-spined sea urchin can be held (although not grabbed), and no sea urchin throws its spines or leaps onto helpless passersby, impaling them.

The skeleton of a sea urchin, known as the test, is a rigid shell made of flat and fused calcareous ossicles (plates). The test is divided into ten sections, extending like slices of an orange between the mouth on the bottom and the anus on top. Five of the sections are pierced with holes through which the tube feet protrude from the water vascular system to the outside.

These sections are the ambulacral plates. Alternating with the ambulacral plates are five plates without holes, the interambulacral areas. When the urchin dies, its spines fall off and the soft body within the test decomposes, leaving the empty test behind.

There are tiny bumps all over the test that fit neatly into concave sockets at the base of the spines. This ball and socket arrangement allows for extensive spine movement. Layers of skin and muscle tissue over the test pull the spines in different directions. Spines usually taper to a point, but the primary spines of the Slate-pencil urchin, *Eucidarias tribuloides,* found from the Carolinas to Florida, are thick and blunt. All urchins possess pedecellarae for cleaning and defense. Some urchins have poison pedecellarae.

Sea Urchins

Urchins eat using a structure, known as Aristotle's lantern, in the mouth. At the center of Aristotle's lantern are five teeth that come together like a bird's beak (except there are five pieces instead of two). These strong teeth allow the urchin to scrape algae off the rocks. As the teeth wear down, they continue to grow. The teeth function as a unit, and can be extended from and pulled back into the mouth.

Arbacia punctulata, the Purple sea urchin, is found from Cape Cod to Florida. In northern New England, the Green sea urchin, *Stronglyocentrosus droebachiensis,* is more prevalent. It is often found in tidepools and below the low-tide line, where it is preyed upon by birds, sea stars, cod, lobsters, and foxes.

The Long-spined sea urchin, common in south Florida, hides in reefs and under ledges during the day. At night it comes out to feed on grass and algae around the reef. Juvenile fishes and small shrimp often hide among the urchin's spines. The long black needlelike spines of the urchin are hollow and brittle, with barbs on the outer surface. When bumped into, the spines break off at skin level. The combined effect of the barbs and the slight toxin on the spines can be quite painful.

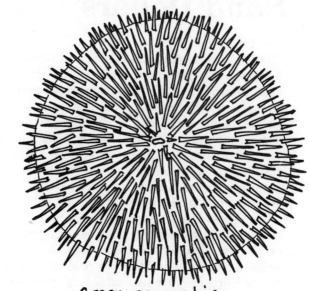

Green sea urchin
Stronglyocentrosus drobachiensis
size: to 3 inches

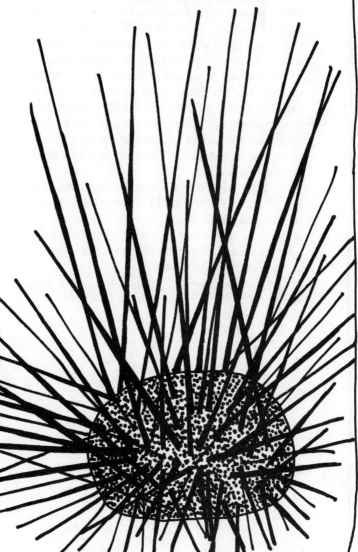

Long-spined sea urchin
Diadema antillarum
size: test to 4 inches
    spines to 1 foot or more

Sand Dollars

Sand dollars are flat versions of sea urchins. Their skeletal plates are also fused to form a rigid internal test. Tiny movable spines cover the test, giving the sand dollar a smooth felt-like texture. Live species of *Echinarachinus parma,* the sand dollar found from Long Island Sound northward, are a reddish-brown to purple collar. The Keyhole sand dollar, *Mellita quinquiesperforata,* found primarily from Cape Hatteras to Florida, is a yellow-brown color.

The short spines of sand dollars are used for burrowing into the sand. The five-petal pattern on the aboral surface corresponds to the five arms of a sea star and the five rows of tube feet of a sea urchin. Tube feet extending from holes in these petals on the sand dollars are used for respiration.

Sand dollars live only on sandy bottoms, where they burrow for protection from waves and predators and to obtain food. As the sand dollar burrows, tiny food particles fall between the dense spines and are transported to the mouth by mucus-coated cilia and tubefeet.

The mouth of a sand dollar is similar to that of a sea urchin. It contains Aristotle's lantern, but it is smaller than the urchin's. When a sand dollar dries, the spines fall off and the bare test eventually bleaches white. The rattling produced by the test is the loosened and dried Aristotle's lantern.

As with most other echinoderms, sand dollars are either male or female, although there are no external differences between the sexes. Eggs and sperm are released into the water through four small openings around the madreporite in the top center of the test.

Despite their spiny, crunchy test, sand dollars are devoured by fish, especially flounder, cod, and haddock.

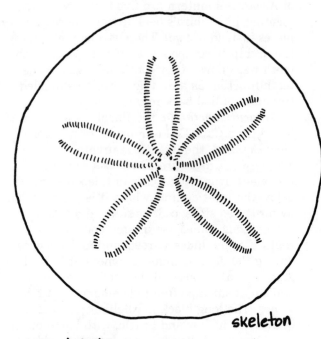

skeleton

Sand dollar
Echinarachnius parma
size: to 3 inches

Keyhole sand dollar
Mellita quinquiesperforata
size: to 3 inches

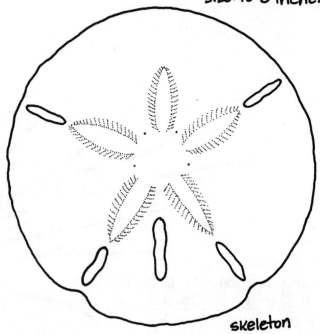

skeleton

Sea Biscuits and Heart Urchins

Brown sea biscuit
Clypeaster rosaceus
size: to 5 inches

skeleton

Both the sea biscuit, *Clypeaster rosaceus,* and the Heart urchin, *Moira atropos,* are relatives of the sand dollar, and both range from the Carolinas to Florida. These species, however, are not flat. The oral surface is flat, but the aboral surface is convex. Their tests are covered with short dense spines used in locomotion. Tube feet extending from the petal-shaped ambulacral grooves are used in respiration. Both have pedecellaria, as do all urchins and sand dollars.

Sea biscuits sit on top of the sand and cover themselves with algae, grass, shells, and other debris. Using their spines, Sea biscuits methodically crawl over the bottom, eating detritus in the sand. Food particles are picked up by the tube feet on the oral surface and brought to the mouth. The anus and madreporite are located at the back edge of the Sea biscuit instead of at the top center.

Heart urchins burrow in the sand with their front edge pointed downward. Mucus-producing tube feet move the sand and maintain a ventilating shaft through the sand on the top side of the Heart urchin. The test of the Heart urchin is fragile and has five deep petaled grooves. The fifth groove helps keep the funnel open. Unlike its relatives, the Heart urchin has its mouth near the front of the oral surface, not in the middle. The anus and madreporite are at the back edge of the Heart urchin.

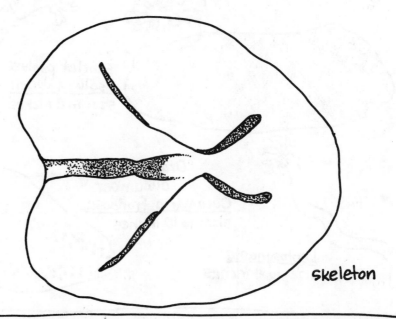

Heart urchin
Moira atropos
size: to 2 inches

skeleton

Sea Cucumbers

Superficially, sea cucumbers do not resemble other echinoderms. They do not have arms or spines. Instead of lying with their mouth on the substrate, they lie on their underside with their mouth forward and their anus at the opposite end. Although their five-part symmetry is not apparent from the side, five parts do radiate from either end. Widely scattered microscopic ossicles embedded in the thick body of the sea cucumber make up the skeleton.

Not all sides of a sea cucumber look alike. Tube feet on the back and sides of the Florida sea cucumber, *Holothuria floridana,* are reduced to mere bumps. The most common species in New England, *Cucumaria frondrosa,* the Orange-footed sea cucumber, has tube feet on all surfaces, but only the ventral ones have well developed suckers. *Psolus fabricii,* another New England cucumber, has tube feet only on its ventral side. The Hairy cucumber, *Sclerodactyla briareus,* found from Cape Code to the Gulf of Mexico, is nearly covered with slender tube feet.

The worm-like *Leptosynapta,* unofficially referred to as the "snot" cucumber, has no tube feet. Sea cucumbers use their feet more for attachment than locomotion.

The ten to thirty modified tube feet that form tentacles around the mouth can be completely retracted. Many cucumbers, such as *Cucumaria,* trap bits of food on their tentacles and then stuff them, one at a time, into their mouth to wipe off the food. *Leptosynapta,* found along the whole coast, consumes detritus as it burrows. *Actinopyga agassizii* ingests sediment and digests the edible portions, excreting the remaining sand in neat strands resembling pearl necklaces.

Sea cucumbers breathe with two respiratory trees on either side of their digestive tract. The cloaca contracts and pumps water through the trees. The Pearlfish, *Carapus,* lives in the respiratory trees of Agassizzi's sea cucumber by day and leaves at night in search of food.

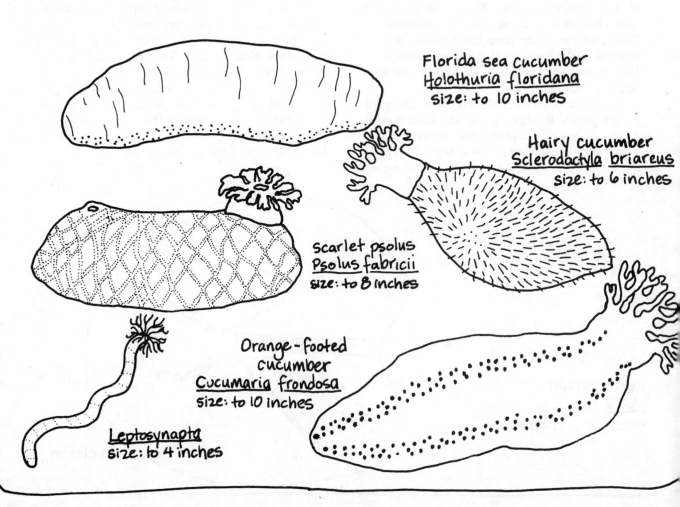

Florida sea cucumber
Holothuria floridana
size: to 10 inches

Hairy cucumber
Sclerodactyla briareus
size: to 6 inches

Scarlet psolus
Psolus fabricii
size: to 8 inches

Orange-footed
cucumber
Cucumaria frondosa
size: to 10 inches

Leptosynapta
size: to 4 inches

Echinoderm True/False Quiz

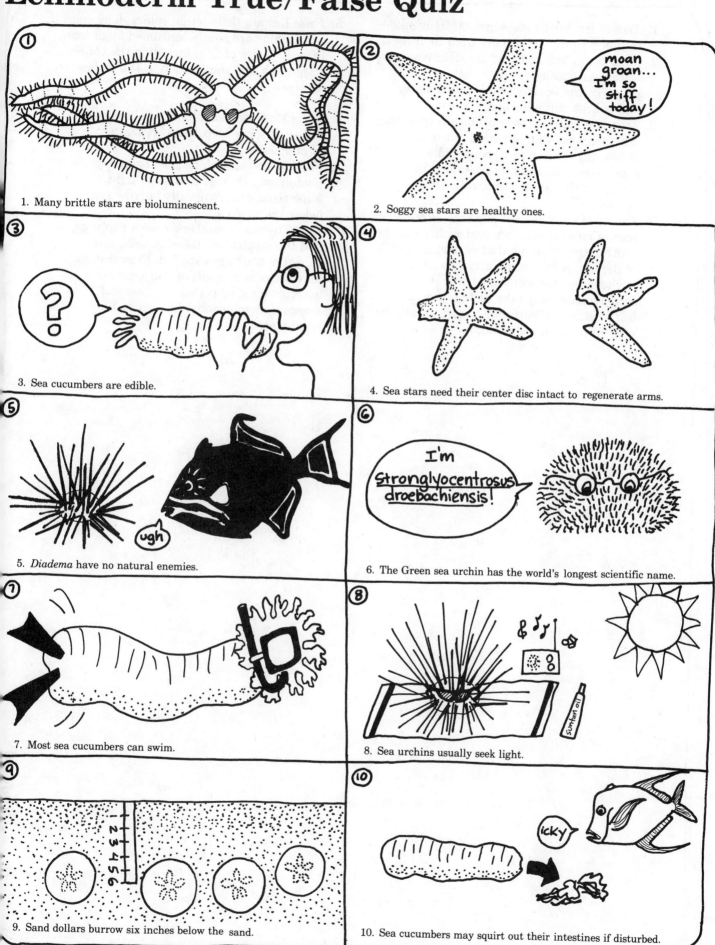

1. Many brittle stars are bioluminescent.

2. Soggy sea stars are healthy ones.

moan groan... I'm so stiff today!

3. Sea cucumbers are edible.

?

4. Sea stars need their center disc intact to regenerate arms.

5. *Diadema* have no natural enemies.

ugh

6. The Green sea urchin has the world's longest scientific name.

I'm Stronglyocentrosus droebachiensis!

7. Most sea cucumbers can swim.

8. Sea urchins usually seek light.

9. Sand dollars burrow six inches below the sand.

10. Sea cucumbers may squirt out their intestines if disturbed.

icky

1. *True* Many brittle stars give off light (like that of a firefly), especially from their legs.
2. *False* Soggy echinoderms are sick. When the water vascular system is functioning properly, the echinoderm is full of water and is kept stiff and rigid.
3. *True* The leathery body of the sea cucumber may not whet your appetite, but in the Orient it is known as *trepang* and is considered a delicacy. Incidentally, the gonads of the Green sea urchin are also edible.
4. *False* Sea stars can usually regenerate any part of an arm and damaged sections of the central disc. It seems that as long as a fifth of the disc is intact and attached to an arm, a whole sea star will grow back. Regeneration may take up to a year. In brittle stars the entire disc must usually be present for arm regeneration.
5. *False* Despite their lethal spines, long spiny sea urchins are avidly consumed by at least fifteen species of fish, including the Queen triggerfish and snails, such as the Helmet shell.
6. *True*
7. *False* Most sea cucumbers can barely crawl, although some deep sea species are swimmers.
8. *False* Most sea urchins are negatively phototropic, that is, they shun light.
9. *False* Sand dollars usually burrow just below the surface.
10. *True* The sea cucumber's cloaca ruptures, and the respiratory trees, gonads, and digestive tract are expelled. Evisceration may occur as a result of foul water or overcrowding, or it may be a normal seasonal phenomenon. Lost parts are regenerated.

13 Tunicates

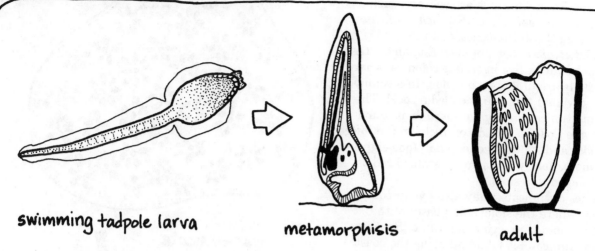

swimming tadpole larva   metamorphisis   adult

Adult tunicates look so much like wads of gum or slabs of blubber that it's hard to believe they are very advanced, complex animals. Who would suspect that vertebrate animals might be descendents of these sessile, immobile blobs so common on rocks and pilings.

Tunicates, also known as sea squirts or ascidians, belong to the phylum Chordata, which also includes fish, birds, mammals, reptiles, and amphibians. At some stage in their life, all chordates have a dorsal nerve cord ("spinal cord"), a supporting notochord ("backbone"), and gill slits.

As larvae, tunicates resemble tadpoles and possess a nerve cord, a notochord, and gill slits. After a short larval life, either as plankton or having been brooded by its parent, the tadpole-like larva settles down on a hard substrate and attaches itself there by means of adhesive organs. It then undergoes metamorphosis, reabsorbing its tail, notochord, and nerve cord. Its internal organs become rearranged, and the tunicate changes into a blob-like creature encased in a leathery or plastic-like sac.

The sac secreted by the tunicate is known as the tunic, and it provides support and protection for the animal inside. Cellulose is a major component of the tunic, although the tunic may contain calcareous spicules as well. The bottom of the tunic has either a rough surface or root-like projections to attach it to the substrate.

Tunicates have two openings: the incurrent and excurrent siphons. They feed on plankton and detritus filtered from the water. Cilia on the pharynx beat to create a current. Water is sucked in through the incurrent siphon and is filtered through the pharynx, which is coated with mucus. An organ, the endostyle, at the side of the pharynx rolls the sheet of mucus with food particles trapped in it and sends it to the esophagus and stomach. Water is then passed through the gill clefts, where oxygen is extracted, and finally exits through the excurrent siphon.

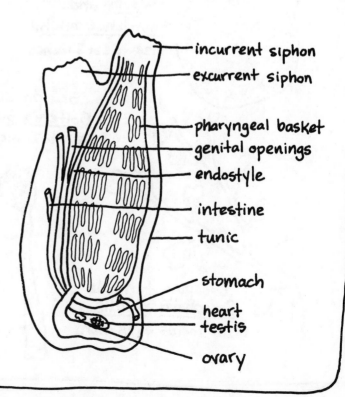

incurrent siphon
excurrent siphon
pharyngeal basket
genital openings
endostyle
intestine
tunic
stomach
heart
testis
ovary

Tunicates

Many species of tunicates are solitary, such as the Sea grape, *Molgula*, or *Bostrichobranchus*, found along the entire coast, and the Sea potato, *Boltenia ovifera*, of New England. Other species, however, live in simple colonies, where individuals are united by a stolon. In a colony, each individual tunicate is called a zooid. The Creeping tunicate, *Perophora viridis*, is found from Cape Code to Florida on rocks in the lower intertidal and subtidal zone. Small (one-eighth of an inch), nearly transparent stalked zooids live attached to a stolon.

Species of tunicates may also live in complex colonies where the stolons and bases of the zooids are joined in a common tunic. *Botryllus*, the Star tunicate, is found all along the coast. It is a complex tunicate that grows as a rubbery sheet on pilings and rocks. Zooids are organized in star-shaped clusters. Each zooid has its own incurrent siphon but shares a common excurrent siphon with neighboring zooids. Sea pork, *Amaroucium stellatum*, which grows as hard, pink rubbery blobs, and the White crust, *Didemnum*, are both complex tunicates found along the entire coast.

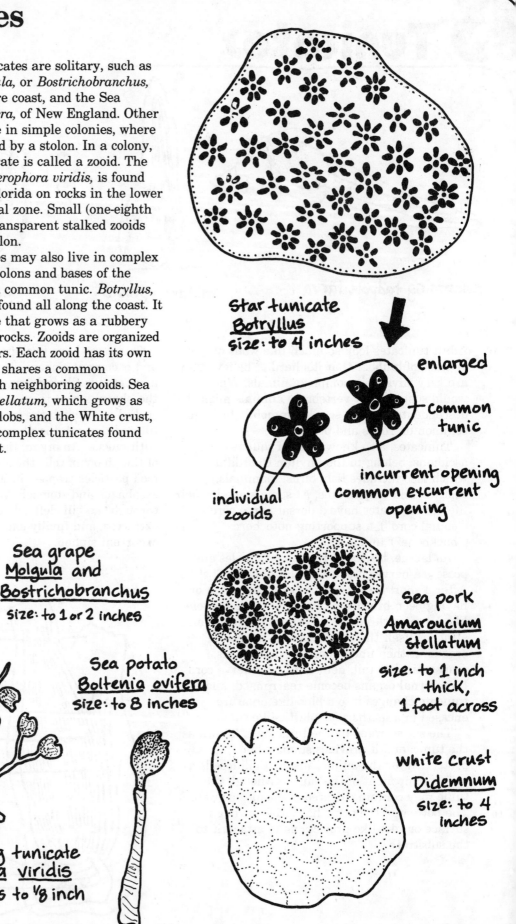

Star tunicate
<u>Botryllus</u>
size: to 4 inches

enlarged

common tunic

incurrent opening
common excurrent opening

individual zooids

Sea grape
<u>Molgula</u> and
<u>Bostrichobranchus</u>
size: to 1 or 2 inches

Sea potato
<u>Boltenia ovifera</u>
size: to 8 inches

Sea pork
<u>Amaroucium stellatum</u>
size: to 1 inch thick, 1 foot across

white crust
<u>Didemnum</u>
size: to 4 inches

Creeping tunicate
<u>Perophora viridis</u>
size: zooids to 1/8 inch

Tunicate True/False Quiz

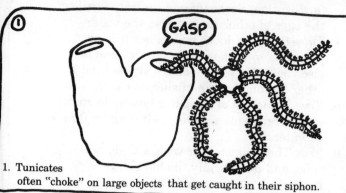

1. Tunicates often "choke" on large objects that get caught in their siphon.

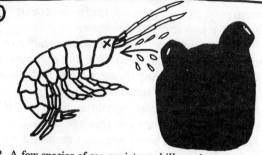

2. A few species of sea squirt can kill passing prey with a quick jet of water.

3. Underwater, tunicates rarely close their incurrent siphons.

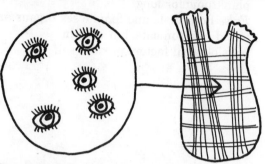

4. Tiny eyespots are scattered on the tunic.

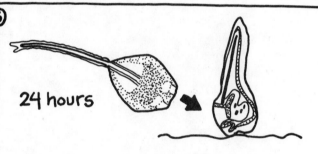

5. Tadpole larvae often settle in less than a day.

6. Tunicates are either male or female.

7. Sea pork is used as a bacon substitute in Iceland.

8. Commensal amphipods are found in tunicates.

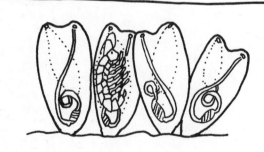

9. Tunicates' hearts can beat in two directions.

10. People are more closely related to tunicates than to lobsters.

1. *False* Tentacles around the interior of the incurrent siphon prevent large objects from being sucked in with the water.
2. *False* Tunicates are not predators; they are exclusively filter feeders. But when removed from the water, they do squirt out a stream of water, thus the name sea squirt.
3. *False* Many tunicates regularly close their siphons. The reasons for this are unclear.
4. *False* No special sense organs are found anywhere on tunicates.
5. *True* Tadpole larvae do not remain planktonic for long.
6. *False* Both male and female sex organs are found in a tunicate, but certain physiological factors prevent self-fertilization. Sperm of one tunicate fertilize the eggs of other tunicates in the water.
7. *False* Sea pork looks edible, but it is not palatable to human beings, in Iceland or elsewhere. However, sharks, skates, and other bottom-dwelling fishes do eat it.
8. *True* Tunicates are popular homes for many small organisms seeking shelter and a flow of water.
9. *True* The tunicate's heart is a U-shaped tube, and periodically the blood flow through it is reversed.
10. *True* People and tunicates both belong to the phylum Chordata. Lobsters are invertebrates in the phylum Arthropodia.

14 Fish

Fish are aquatic, cold-blooded vertebrates that breathe with gills and have fins instead of limbs. Nearly half of the 40,000 known species of vertebrates are fish. Ninety percent of these fish species belong to the class Osteichthyes, the Bony fish. Most of the other 10 percent belong to the class Chondrichthyes, the Cartilaginous fish. Bony fish have a skeleton made of bone; cartilaginous fishes have a skeleton made of cartilage. Sharks, rays, and skates are cartilaginous fish, whereas almost all other fish, such as cod, tuna, and salmon, are bony fish.

Fish differ greatly in breeding habits. Most fish species reproduce through external fertilization. Some fish merely swim next to each other while releasing eggs and sperm into the water. In this case, with egg and sperm randomly discharged into the water, there is a good chance that many eggs will never be fertilized. To compensate for this, the female produces thousands, even millions, of eggs, thus ensuring that sperm cannot help but bump into some of them. A few species, especially the sharks, display internal fertilization. Male sharks actually place sperm into the reproductive tract of the female.

Although some fish attach their fertilized eggs to the bottom and others build nests, most species of bony fish start life as tiny planktonic eggs, adrift in the sea. The eggs float (some contain a drop of oil to increase their buoyancy) and disperse from the breeding ground.

Larvae develop inside the eggs, nourished by the yolk. Depending on the species, they hatch out days or weeks later. Fish larvae have large eyes, a short trunk, a long tail, and a transparent fin extending from head to tail. Between their head and anus is a glob of yolk that sustains the young fish until its muscles and jaws are well enough developed so it can fend for itself. When the yolk is depleted, the larvae feed on plankton. Eventually the fish larvae develop into fry (young fish), and then into adults. Some fish mature in weeks, while others take years to reach sexual maturity. Naturally, only a small percentage of fish larvae live to adulthood.

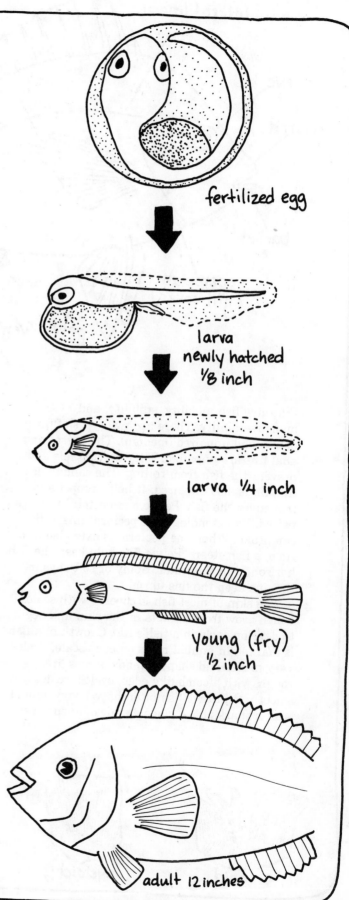

fertilized egg

larva
newly hatched
⅛ inch

larva ¼ inch

young (fry)
½ inch

adult 12 inches

The External Bony Fish

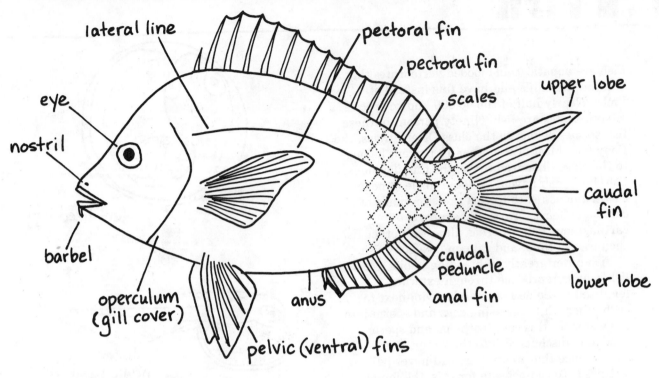

lateral line

pectoral fin

pectoral fin

upper lobe

eye

scales

nostril

caudal fin

barbel

caudal peduncle

lower lobe

operculum (gill cover)

anus

anal fin

pelvic (ventral) fins

Fins give fish mobility, stability, and maneuverability. There are two types of fins: paired and unpaired (median). The dorsal and anal fins are unpaired. They act as keels and prevent the fish from rolling. The caudal, or tail, fin is also unpaired. It helps propel and maneuver the fish. Pelvic and pectoral fins are paired. By extending both pectoral fins, a fish can brake. When one pectoral is extended at a time, a fish steers. Pelvic fins help keep the fish horizontal in the water. Spiny or soft rays in the fins keep the fins upright.

The skin of most fish is covered with scales, which grow from pockets in the skin and begin to develop early in a fish's life. Growth of a fish is marked in annual rings on each scale. Scales vary in size and shape. Ctenoid scales are rough, with a comb-like edge; cycloid scales are smooth. Primitive fish often have heavy ganoid scales; sharks have placoid scales. Mucus over the skin and scales protects the fish from disease.

The color of fish is due to a combination of chromatophores (pigment cells) and iridocytes (reflective cells). These cells can contract or expand to alter the color of the fish. Many fish exhibit countershading: dark on top, where the sun hits, and light underneath, which is usually shaded. This helps them to blend in with the sky (when the fish is viewed from below) and the sea bottom (when viewed from above).

Other external anatomical features include the mouth, which may contain teeth; the barbel, used in tasting and feeling; the nostrils, and the bony gill cover, also called the operculum. The lateral line is made up of sensory cells that detect disturbances in the water. Some fish have a faint lateral line, while others have none at all.

ctenoid

cycloid

ganoid

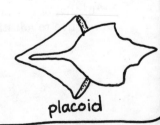

placoid

The Internal Bony Fish

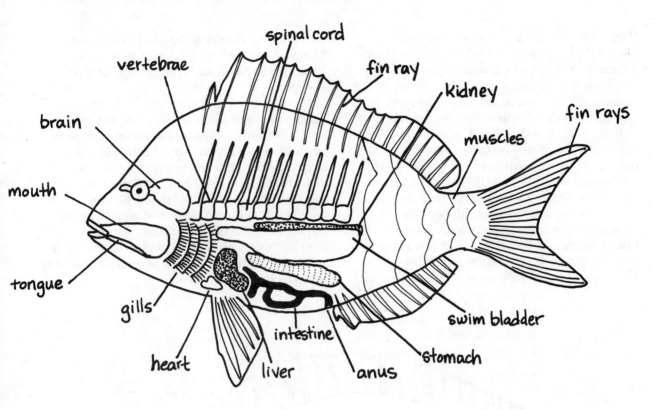

Bony fish have a skeleton of bone. Cartilaginous fish (sharks, rays, and skates) have a skeleton of cartilage. The skeleton of bony fish is similar to that of other vertebrates. The vertebral column starts from the skull and ends where the tail begins. Fish bones are strong, yet light weight. As with other creatures, the skeleton is a site of muscle attachment.

Fish swim by flexing their streamlined bodies. Waves of contractions travel down successive muscle segments along the fish. These waves push the water backward, causing the fish to move forward. Fins assist in braking, steering and maneuvering. Each fin is moved by its own set of muscles.

Most fish have a swim bladder. This is a gas-filled sac that the fish inflates or deflates in order to maintain *neutral buoyancy* in the water. When a fish is neutrally buoyant, it neither sinks to the bottom nor floats to the surface, but is able to "hang" in the water. Gas for the swim bladder is manufactured in a gland. Some fish have a swim bladder with an opening. To deflate the bladder, gas is expelled through the opening and exits through the mouth and gills. Fish with closed bladders expel excess gas into their blood. Sharks, skates, rays, clingfish, and flatfish are examples of fish that do not have swim bladders.

By breathing, organisms take in oxygen, which is used in bodily reactions that release the energy needed to live and grow. Fish breathe by taking water in through their mouth. As they close their mouth, water is forced over their gills and out the operculum. Fish gills are made of fine filaments attached to a flexible skeletal arch. Blood circulates through these fine filaments, and hemoglobin in the blood absorbs the dissolved oxygen in the water. Water exiting through a fish's gills contains 80 percent less oxygen than when it entered through the mouth. Some fish have gill rakers (screen-like structures) across their gills. The gill rakers filter out food and debris from the water as it passes over the gills.

The Cod and the Summer Flounder

Cod, *Gadus callarias,* are one of the most plentiful and important food fish in the North Atlantic. (Young cod are sold as scrod.) A cod weighing 211 pounds was caught off the Massachusetts coast in 1895, but the more typical size is twelve to thirty-five pounds.

Cod are bottom-dwelling fish, usually found within six feet of the bottom. There are small teeth in both jaws. Although they prefer large mollusks (which they swallow whole, shell and all), Cod are not finicky eaters. They feed on a variety of sea creatures including crabs, sea stars, sea cucumbers, sea urchins, worms, tunicates, skates, hydroids, squid, sea horses, and algae. Chunks of leather and rubber have also been found in their stomach. In turn, Cod are eaten by sharks and pollock, among other things. Adult Cod prefer water temperatures of 32–55°F and are found from Maine to New Jersey.

Flatfish, such as Yellowtail, Halibut, Hogchoker, and the flounders, are not really flat. It is more accurate to say that they swim on their side. As young fry, flatfish look like any other fish, but as they mature one eye migrates around their head and ends up next to the other eye. Adult flatfish are bottom dwellers, living with their eyeless side against the bottom. The edible Summer flounder, *Paralichthys dentatus,* is found from Maine to South Carolina. It can change colors to blend in with the bottom.

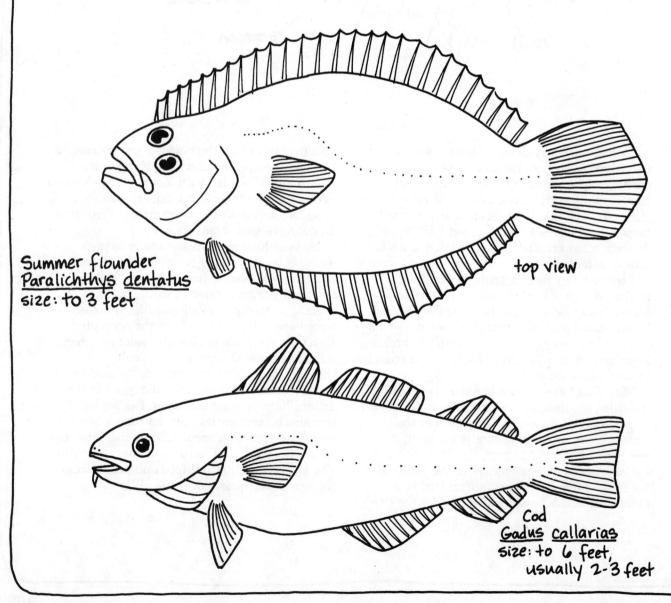

Summer flounder
Paralichthys dentatus
size: to 3 feet

top view

Cod
Gadus callarias
size: to 6 feet,
usually 2-3 feet

Remoras and Sea Horses

Remoras, also called Shark suckers, have a highly modified first dorsal fin that acts as a sucking disc. Typically, Remoras are found hitching a ride on sharks, but they also attach themselves to dolphins, swordfish, ships, turtles and skin divers. Although a few Remoras have wandered or been carried as far north as the Gulf of Maine, they are primarily tropical and subtropical.

The sucking disc on the Remora's head is flat, with cartilaginous cross plates and raised edges. Once the Remora attaches itself to its host, it can release itself at any time merely by swimming forward. Forward motion of the host, however, will not dislodge the Remora. Remoras usually cling to the side of their host, but they have also been found in the gill cavity and mouth of large fish.

Remoras are not parasites. They do not harm the shark. In fact, they may help it by eating parasites on its skin. It was originally thought that Remoras shared the food of their host, but more than likely they are just hitching a ride and will detach themselves to swim after food themselves.

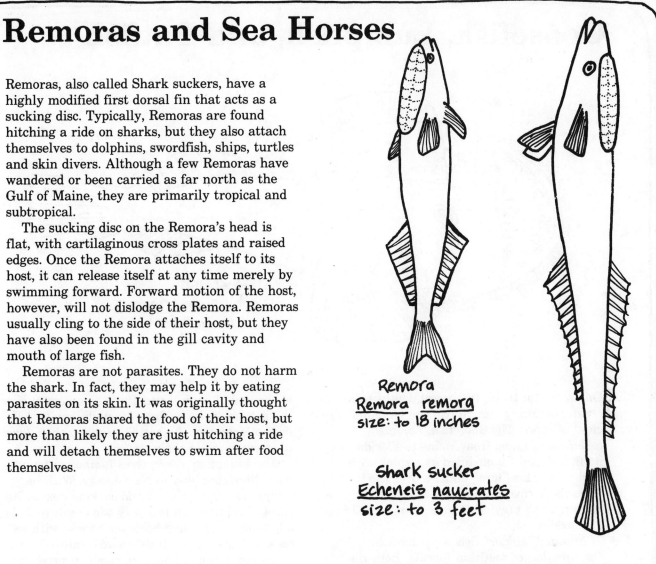

Remora
Remora remora
size: to 18 inches

Shark sucker
Echeneis naucrates
size: to 3 feet

Seahorse
Hippocampus
size: to 6 inches

Swimming is not the strong suit of the sea horse. Lacking pelvic and caudal fins, a sea horse swims upright, using its ear-like pectoral fins with its dorsal fin for propulsion. This technique is fine for gliding, but not for contending with strong currents. Because they are such weak swimmers, sea horses spend much of their time with their prehensile tail wrapped around seaweed and seagrass. When small creatures go past, sea horses suck them into their mouths.

The scales of sea horses are fused together, forming interlocking armored plates that give the sea horse that spiny, ridgy, knobby look.

The male sea horse deposits sperm on the eggs as the female expels them. Then the female deposits the fertilized eggs in a pouch on the male's abdomen. The eggs are brooded there until they hatch. Sea horses are found from New Hampshire to Florida.

Goosefish, Lumpfish, and Batfish

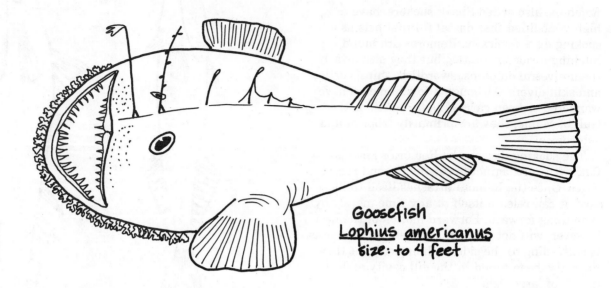

Goosefish
Lophius americanus
size: to 4 feet

One need not travel to abyssal depths to see odd creatures. Many bizarre fish can be found very close to shore. The Goosefish, *Lophius americanus,* found from Maine to Florida, has its first dorsal fin modified as a spine with a fleshy tip. This "lure" dangles over the Goosefish's cavernous mouth and attracts unsuspecting prey, which are then sucked in and swallowed.

Another "angler" fish is the batfish, *Ogocephalus,* of southern Florida. Both the Goosefish and batfish have fleshy pectoral fins at the end of bony joints. Batfish can crawl over the bottom on these fins, although they usually lie motionless on the bottom, angling their lure to attract prey to their mouth.

The Lumpfish, *Cyclopterus lumpus,* is found from Newfoundland to New Jersey. With its lumpy body, tiny mouth and sucking disc on its chest, the Lumpfish is a very odd creature. It is a bottom dweller and holds onto rocks with its sucker. Occasionally it drifts with rafts of rockweed. Lumpfish feed on small invertebrates.

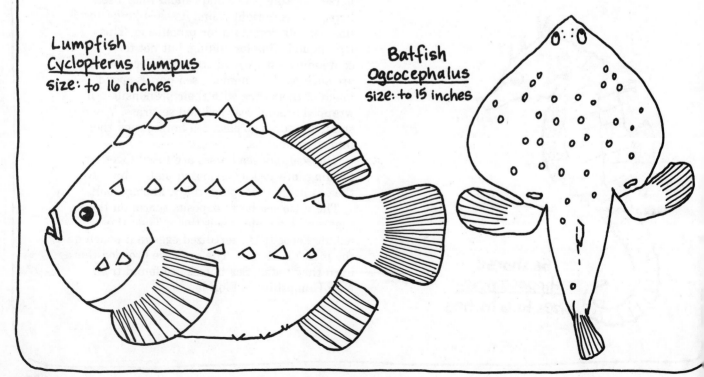

Lumpfish
Cyclopterus lumpus
size: to 16 inches

Batfish
Ogocephalus
size: to 15 inches

Minnows

The Common mummichog, *Fundulus heteroclitis,* is variously known as the chub, killifish or salt water minnow. It is found from Maine to Florida, and sticks close to shore, preferring harbors, tidal creeks, tidepools and estuaries. This is a very adaptable fish, with a high tolerance for changes in oxygen availability, salinity, and temperature. The Common mummichog is omnivorous. It eats living and dead plants and animals and has an affinity for mosquito larvae.

Another fish tolerant of changing oxygen levels, temperature, and salinity is the Sheepshead minnow, *Cyprinodon variegatus,* found from Cape Cod to Florida. This feisty fish has formidable teeth and is known to eat fish larger then itself. It is often used as bait.

The Striped killifish (or Striped mummichog), *Fundulus majalis,* is also omnivorous. It lives in shallow water from Cape Cod to Florida and is most common along open beaches rather than in tidal creeks or harbors. If stranded by the receding tide, this acrobatic fish can flop its way back into the water.

The Atlantic silverside, *Menidia menidia,* is a slender fish covered with large scales. Silversides are found from Nova Scotia to Florida and are eaten by many food fish, including bluefish, bass, and mackerel. They live close to shore and often congregate in large schools. Schools are made up of one kind of fish, all of which are about the same size. Members of the school do the same thing at the same time, following the leader. When the group turns, whoever is in front leads. Good eyesight is crucial in keeping the school together, although other sensory receptors, such as the lateral line, are also important. Schooling is safer than swimming alone, and makes it easier to find food and scare predators.

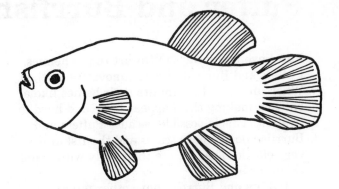

Common mummichog Fundulus heteroclitis
size: to 4 inches

Sheepshead minnow Cyprinodon variegatus
size: to 3 inches

Striped mummichog Fundulus majalis
size: to 6 inches

Silverside Menidia menidia to:
size: to 4 inches

Puffer and Burrfish

Fish have many ways to thwart their enemies. Puffers and Burrfish use an innovative method. When annoyed, they inflate their bellies like a balloon, making them appear large and hard to swallow. This is possible because Puffers and Burrfish do not have scales and their skin is very elastic. They inflate their belly with water or air, or both.

Puffers and Burrfish have other unique anatomical features. They lack pelvic fins, and their gill openings are reduced to small slits. Their teeth are fused together to form plates. Puffers appear to be "buck-toothed" because their jaw plates are divided down the middle. Burrfish have a solid plate (like one big tooth) across their top jaw and one across their bottom jaw.

The Puffer, *Sphaeroides maculatus,* is a slender, normal-looking fish when not inflated. Once it inflates, the Puffer floats belly up at the surface. As soon as it deflates its belly, it swims away underwater. The Puffer's skin is covered

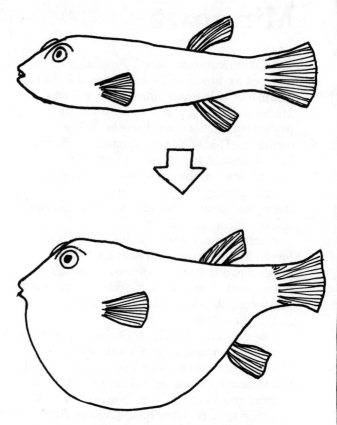

Puffer <u>Sphaeroides</u> <u>maculatus</u>
size: to 10 inches

Burrfish <u>Chilomycterus</u> <u>schoepfii</u>
size: to 10 inches

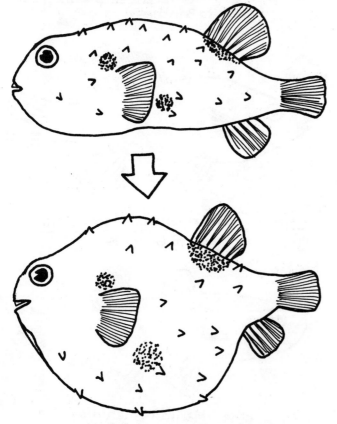

with tiny prickles. The combined effect of the spines and the inflated belly ward off many predators.

Puffers use their teeth to crush and devour mollusks and crustaceans. They can be excellent food, but their viscera (guts) are poisonous, and if improperly prepared, Puffers can be extremely toxic. They are common in shallow water from Cape Cod to Florida.

Burrfish, *Chilomycterus schoepfi,* are covered with large triangular spines. When not inflated, Burrfish resemble loaves of bread. They have a large head and stocky body. Burrfish usually hide in protected spots, such as caves, crevices, and reefs. Like the Puffer, the Burrfish crushes crustaceans and mollusks with its teeth. Their flesh may be poisonous. Burrfish live in inshore waters from Cape Cod to Florida.

Barracuda and Swordfish

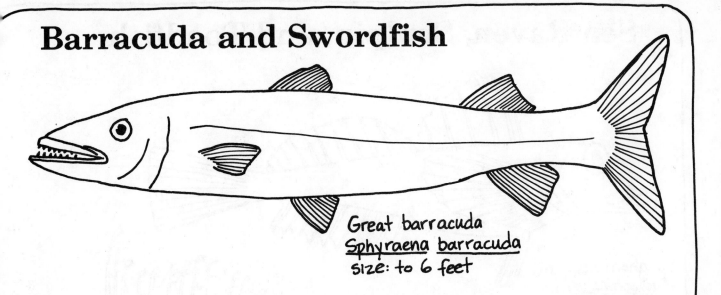

Great barracuda
Sphyraena barracuda
size: to 6 feet

The surest way to be injured by a Great barracuda, *Sphyraena barracuda,* is to eat it. More people have contracted food poisoning eating barracuda than by eating any other fish. And despite the terrifying reputation of the Great barracuda, there are no substantiated records of unprovoked attacks.

Although the Great barracuda's reputation as a man-eater is totally unfounded, the sight of a large adult can still be unnerving. Young barracudas may school, but large ones tend to be solitary and territorial. Barracudas are inquisitive fish and like to keep an eye on snorkelers or divers in their territory. They hang motionless in the water, smiling and staring at passersby. With their jutting lower jaw and mouthful of sharp teeth, a Great barracuda hardly looks friendly.

The Northern barracuda, *Sphyraena borealis,* a totally innocuous relative of the Great barracuda, grows to eighteen inches long and is found from Cape Cod to Florida. Great barracudas live from South Carolina to Florida.

The Swordfish, *Xiphias gladius,* is found from Maine to Florida, usually in the offshore waters of the Gulf Stream. Its long spear-like upper jaw is a formidable weapon, and its deeply forked tail makes it a powerful swimmer.

Although they have no teeth or scales, adult Swordfish feed on bluefish, mackerel and squid. Full grown Swordfish are preyed upon only by the largest carnivores. A 120-pound Swordfish was found whole in the stomach of a 730-pound Mako shark. Swordfish meat is also sought by people.

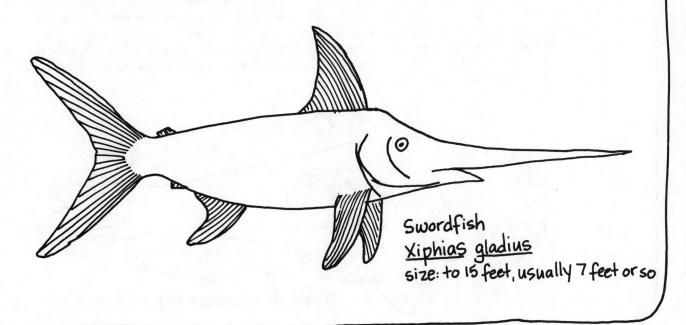

Swordfish
Xiphias gladius
size: to 15 feet, usually 7 feet or so

Sea Raven, Sculpin, and Toadfish

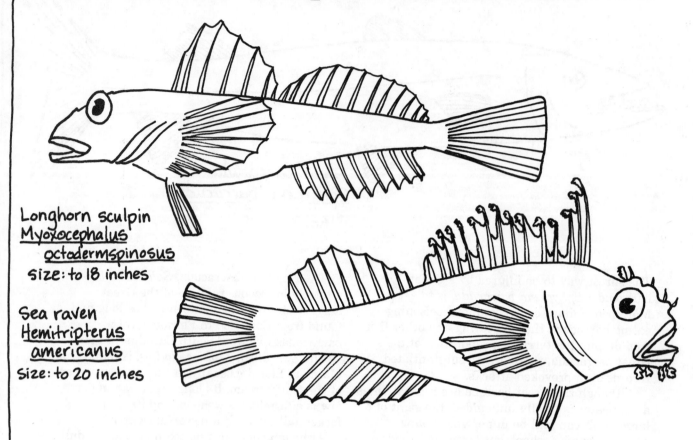

Longhorn sculpin
Myoxocephalus
octodermspinosus
Size: to 18 inches

Sea raven
Hemitripterus
americanus
Size: to 20 inches

Many marine creatures produce sounds, both in and out of the water. The Longhorn sculpin, _Myoxocephalus octodecimspinosus_, Sea raven, _Hemitripterus americanus_, and Toadfish, _Opsanus tau_, all make grunting noises.

The Longhorn sculpin and Sea raven are both found from Maine to New Jersey in shallow water. As with other advanced fish, their first dorsal fin is spiny, and the second one is soft. Sculpin are scavengers and bite at any baited hook. Removing them from a hook is tricky, since sculpin will erect their spiny dorsal fin and fan out their spiny gill covers.

Sea ravens have large teeth and fleshy growths under their mouth. They are prickly all over. When annoyed, Sea ravens can inflate their belly with water. Despite their unappetizing appearance, sculpins and Sea ravens are edible. Sea ravens are often used as lobster bait.

The Toadfish lives from Cape Cod to Florida. This hardy fish can survive out of water for quite a while. Toadfish will eat almost anything. Rarely can any creature escape the Toadfish's huge mouth and big blunt teeth.

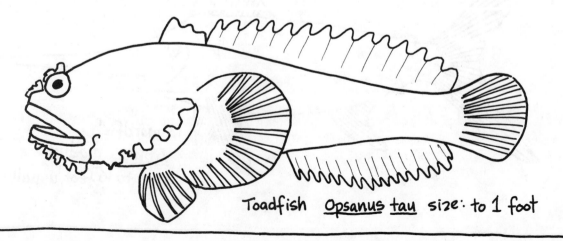

Toadfish   _Opsanus tau_   Size: to 1 foot

Eels and Morays

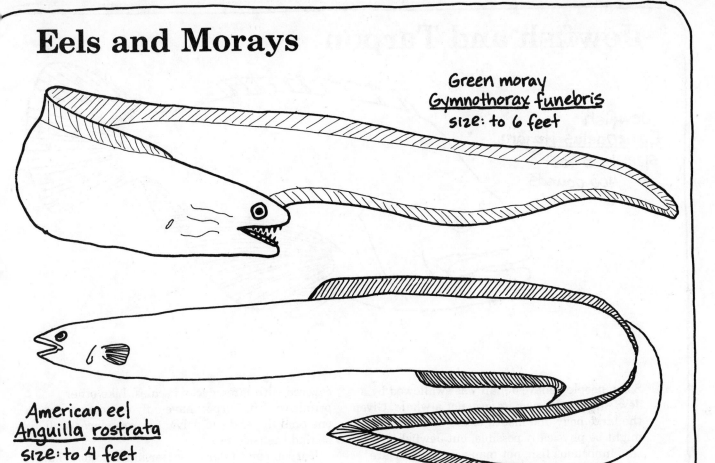

Green moray
Gymnothorax funebris
size: to 6 feet

American eel
Anguilla rostrata
size: to 4 feet

People often mistake eels for sea snakes, but there are no sea snakes in the Atlantic Ocean. Eels are fish with long bodies. Their dorsal, caudal, and tail fins are joined and run as one long fin down their back and up their belly. There are no pelvic fins. Eels have tiny scales or no scales at all.

The American eel, *Anguilla rostrata*, lives all along the coast from Maine to Florida. This eel is a scavenger and will eat practically anything, including the bait on a hook. American eels are chiefly nocturnal and may pass the day buried in mud, waiting to scavenge at night.

The life cycle of the American eel remained a mystery until recently. Mature eels travel to the Sargasso Sea (page 36) off the coast of the Bahamas to breed. Fertilized eggs hatch into small, flat, transparent larvae. The larvae make the long trek back to North America and look more eel-like by the time they reach the coast. Some of the eels then ascend rivers and head to fresh water, while others stay in tidal creeks, harbors, or estuaries. Several years later, at

sexual maturity, eels head out to sea to spawn. This is the opposite of salmon, shad, and alewife, which mature at sea and spawn in fresh water. American eels die after spawning. Europeans eat quite a bit of eel, but Americans are less adventurous.

The Green moray eel, *Gymnothorax funebris*, lives in tropical waters around the world. Here it is found as far north as New Jersey. Green morays have leathery, scaleless blue-grey skin. Their green color is due to a yellowish mucus that completely covers their blue skin. One reason morays look so eerie is that they continually open and close their mouth to pump water over their gills. Water exits through tubular openings near their head.

Green morays hide in caves and reefs by day. Unless provoked, they are harmless. Their teeth are sharp, but not poisonous. Decaying, contaminated food particles on them, however, may cause a seriously infected bite. Many species of moray cause food poisoning if eaten.

Jewfish and Tarpon

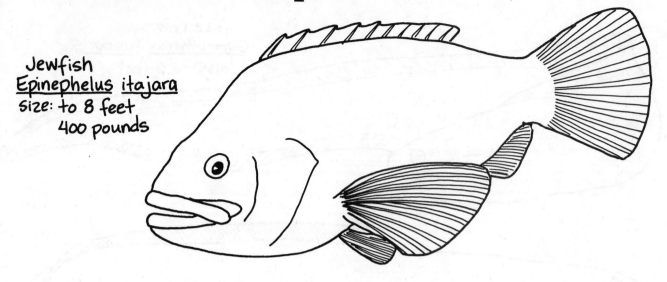

Jewfish
Epinephelus itajara
size: to 8 feet
400 pounds

Some people believe Jonah was swallowed by a Jewfish, *Epinephelus itajara,* not a whale. Given the large body and mouth of the Jewfish, that might be physically possible, but Jewfish are shy, innocuous fish, not man-eaters. They are members of the Sea bass family and are considered a type of grouper. Jewfish are good to eat, but large ones (they can reach 700 pounds!) may cause food poisoning.

Jewfish are bottom dwellers. All groupers, including the Jewfish, first mature as females. After producing eggs, they gradually change sex and become males. Jewfish are found in southern Florida, especially around reefs.

Tarpon, *Megalops atlantica,* resemble gigantic silversides. They are bright silver and are covered with large cycloid scales. Like other primitive fish, Tarpon have only one dorsal and one anal fin, and the pelvic fins are placed well behind the pectorals.

Tarpon can't tolerate water much colder than 65°F, and are killed by sudden drops in water temperature. They live from Long Island to Florida in relatively shallow water. Occasionally they get as far north as Cape Cod. They can live in fresh or salt water.

Despite their well-developed gill, Tarpon must come to the surface to breathe air. They feed on crustaceans and fish, such as mullet. They, in turn, are eaten by sharks, dolphins, and porpoises. Tarpon are not edible by human beings.

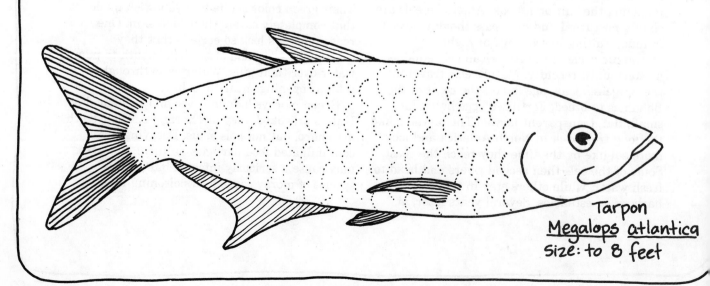

Tarpon
Megalops atlantica
size: to 8 feet

Tropical Reef Fishes

Most tropical fish live on or around healthy coral reefs. In the continental United States, the only coral reefs are found off the coast of the Florida Keys. The caves, cracks, and crevices in the reef provide hiding places for the thousands of organisms that live there.

Tropical reef fish are incredibly diverse in color, shape, and habits. Most are slim and can dart easily through small passageways. The reason for their bright colors is less obvious. One avid fish-watcher has concluded that it is "beauty beyond practicality." The colors usually call attention to the fish, which may help members of the same species recognize one another. Bright colors also serve as a territorial warning to other fish. Many fish display disruptive coloration, that is, their outline is broken up with patches, spots, or bands. Other fish have a stripe through their eye. Predators usually aim for the eye of their victim, but the stripe helps conceal it.

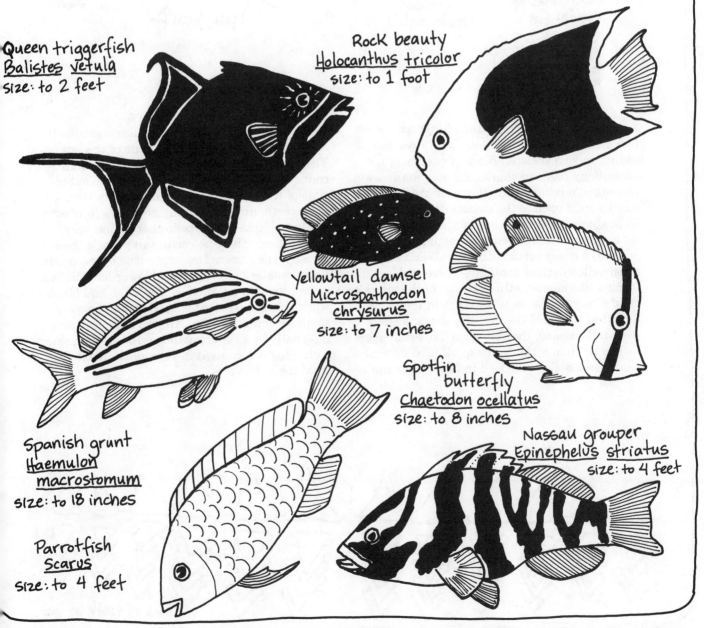

Queen triggerfish
<u>Balistes</u> <u>vetula</u>
size: to 2 feet

Rock beauty
<u>Holocanthus</u> <u>tricolor</u>
size: to 1 foot

Yellowtail damsel
<u>Microspathodon</u>
<u>chrysurus</u>
size: to 7 inches

Spotfin butterfly
<u>Chaetodon</u> <u>ocellatus</u>
size: to 8 inches

Spanish grunt
<u>Haemulon</u>
<u>macrostomum</u>
size: to 18 inches

Nassau grouper
<u>Epinephelus</u> <u>striatus</u>
size: to 4 feet

Parrotfish
<u>Scarus</u>
size: to 4 feet

The External Shark

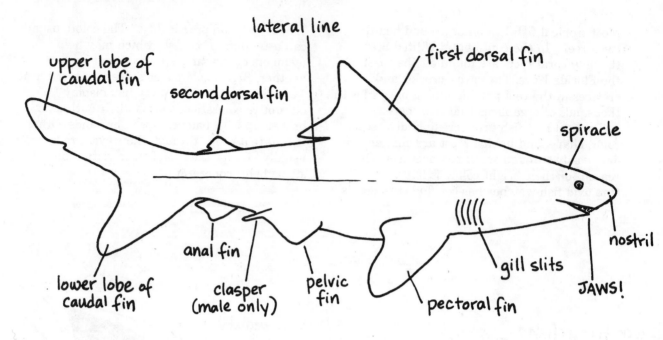

upper lobe of caudal fin

lateral line

second dorsal fin

first dorsal fin

spiracle

nostril

JAWS!

gill slits

pectoral fin

pelvic fin

clasper (male only)

anal fin

lower lobe of caudal fin

Approximately 300 out of the 20,000 species of fish are sharks. Sharks, along with rays and skates, have a skeleton made of cartilage, not bone. Bony fish and sharks differ in many ways. Instead of having a bony flap over the gills, sharks' gills open to the outside through slits. Most sharks have five slits to a side; other have up to seven. The spiracle is a vestigial gill slit.

Sharks are covered with tiny placoid scales, often called dermal denticles. These scales, formed like small teeth, grow out of the shark's skin. Many of the denticles point in the same direction—toward the tail. If a shark is rubbed from tail to head, the skin feels like sandpaper.

The teeth of a shark are specialized dermal denticles loosely embedded in tissue over the jaw. Three to fifteen sets of replacement teeth are neatly arranged on the inside of the jaw. They point downward and only turn up in a conveyor belt system as they reach the outer edge of the mouth. Sharks continually shed their teeth, always growing new ones in reserve.

Most sharks are torpedo-shaped and very streamlined. Their powerful tail propels them through the water. The upper lobe of the caudal fin is usually larger than the lower lobe. Strong-swimming sharks, such as the Mako and Great white, have a crescent shaped tail and a very small second dorsal fin. All fins are supported internally by rays of cartilage. In male sharks, a clasping device used in reproduction extends from the pelvic fins.

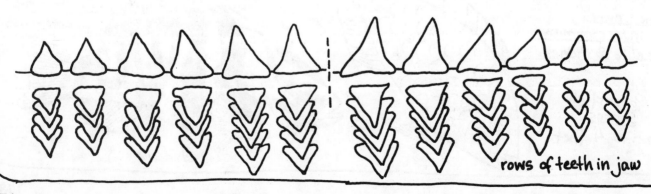

rows of teeth in jaw

The Internal Shark

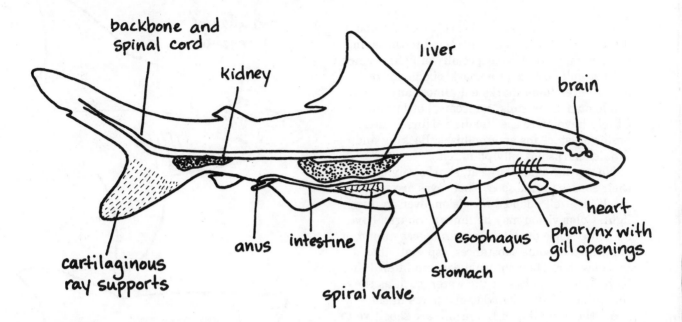

backbone and spinal cord

kidney

liver

brain

cartilaginous ray supports

anus

intestine

spiral valve

stomach

esophagus

pharynx with gill openings

heart

A shark has no bones. Its skull and vertebrae are made of cartilage. Cartilage is relatively flexible, but the backbone may be calcified for extra rigidity. Unlike the tail of bony fish, which starts where the backbone ends, a shark's backbone extends well into the tail, making the tail very powerful. There is no rib cage to protect a shark's internal organs, so despite their tough exterior, sharks can't tolerate much abuse, especially out of the water. They are held together only by skin and muscle.

Sharks have no swim bladder. Instead, they use their liver, which can be more than 15 percent of their body weight, to control buoyancy. In addition to producing bile, a shark's liver serves as a fat and oil reserve. The oil helps keep the shark from sinking to the bottom. Unlike bony fish, which can change the size of their swim bladder, a shark cannot readily change the size or buoyancy of its liver.

Water enters a shark through the mouth, passes over the gills, where oxygen is extracted, and then leaves the body through the gill slits. Many sharks must keep swimming to keep breathing. They are not able to pump water over their gills, and therefore cannot hang motionless in the water without becoming short of breath. Some sharks, especially those that live inshore, have muscles around their gills, enabling them to rest and breathe at the same time.

During copulation, a male shark inserts his clasper (on the pelvic fin) into the female. Sperm passes from the male along a groove in the clasper and into the female. Some species of sharks lay eggs in leathery cases, and the babies hatch in the water months later. Others retain the fertilized eggs in their bodies, and baby sharks emerge live from the female after hatching inside her. A few species actually have a placental attachment to the developing embryos. No sharks care for their babies after hatching.

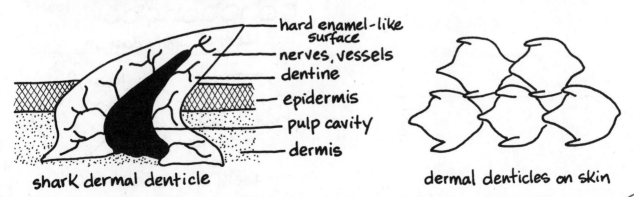

hard enamel-like surface

nerves, vessels

dentine

epidermis

pulp cavity

dermis

shark dermal denticle

dermal denticles on skin

Shark Senses and Feeding Behavior

Sharks are scavengers and carnivores. Although it seems most sharks are attracted to thrashing movements and blood (especially of fish), we are not yet able to predict a shark's behavior or food choice. Some sharks eat almost anything, edible or not: tin cans, raincoats, bottles, etc. Others have more specific diets. Nurse sharks prefer crustaceans and mollusks. Basking and Whale sharks eat only plankton.

No matter what they eat, sharks rely on their senses to find food. Their eyes are usually small but well adapted for vision, even in dim light. Color vision may be limited. Sharks have an acute sense of smell and can detect minute quantities of some substances, especially blood, in the water. Their internal ear and lateral line both detect impulses in the water, such as the thrashings of sick fish. Sac-like pores in the head, the ampullae of Lorenzini, are sensitive to electric impulses, especially the electrical signals created by the muscular movements of thrashing animals. All these senses make sharks efficient, astute predators.

Because a shark's mouth is located on the underside, it had been thought that a shark had to roll over to bite and grab its prey. That is not the case. A shark's jaws are very flexible and loosely set in the mouth. When a shark grabs its prey, it lifts its snout and thrusts out its jaws, without having to get above or to the side of the prey.

Each species of shark has its own type of teeth. Differences in tooth structure reflect differences in their function and in the shark's diet. The relatively flat teeth of Nurse sharks crush mollusk shells. Bladelike, triangular teeth of Great white sharks cut flesh from large prey.

With the variety of items sharks eat, it's no wonder people are occasionally bitten. But shark attacks are very rare. Most occur when a person provokes a shark or if the person is bleeding profusely in the water, such as after an air or sea disaster.

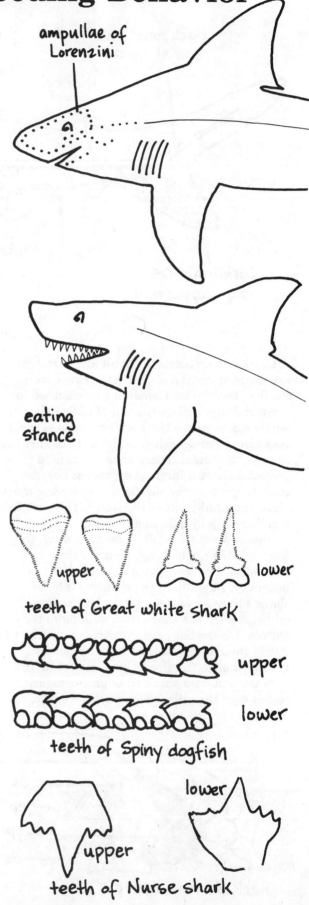

ampullae of Lorenzini

eating stance

teeth of Great white shark

upper

lower

teeth of Spiny dogfish

upper

lower

teeth of Nurse shark

upper

lower

Dogfish

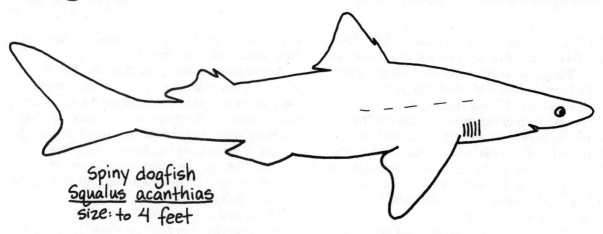

Spiny dogfish
Squalus acanthias
size: to 4 feet

Not all sharks are of the behemoth *JAWS* variety. Dogfish are small sharks, very common along the Atlantic Coast. The Spiny dogfish, *Squalus acanthias,* found from Maine to North Carolina, is by far the most abundant shark in New England. Along the front edge of each dorsal fin lies a sharp spine used for defense.

The voracious eating habits of the Spiny dogfish have long put it at odds with commercial fishermen. Being very gregarious, Spiny dogfish swim in schools and drive away mackerel, herring, cod, and haddock, preying on many of them. Before the days of bottom trawling, dogfish actually prevented fishing in parts of the Gulf of Maine because they took bait and got hooked themselves. When fresh, the Spiny dogfish is quite delectable (as are most sharks), but their reputation as a menace has overshadowed their palatability. Spiny dogfish are not discriminating eaters, and will devour any fish smaller than themselves, as well as worms, squid, and comb jellies.

The Smooth dogfish, *Mustelus canis,* lives in coastal waters from Florida to Cape Cod. Its teeth are very small, and several rows function together as a grinding surface. Smooth dogfish have an insatiable appetite for lobster and large crabs. They often prey on small fish, eating sick or injured ones. The Smooth dogfish travels in huge schools along the bottom, eating most crustaceans in their path. This dogfish, like the Spiny dogfish, is very edible, but more often than not, when caught, it is beaten on the head and tossed back into the water to die.

Smooth dogfish give birth to live young, and there is a placental attachment between the mother and the embryos.

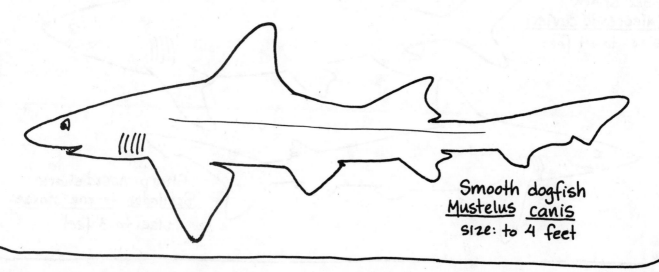

Smooth dogfish
Mustelus canis
size: to 4 feet

Requiem Sharks

The Requiem sharks (Family Carcharhinidae) are streamlined fish with very sharp, blade-like teeth. Most sharks have two eyelids, an upper and a lower, neither of which can move. Requiem sharks have a third eyelid, the nictitating membrane, which comes up from the bottom to cover and protect the shark's eye. The nictitating membrane usually comes up as the shark is feeding so the shark isn't poked in the eye by the prey. Requiem sharks bear live young; some species have a placental attachment with their embryos.

The Brown, or Sand bar, shark, *Carcharhinus milberti,* found from Cape Cod to Florida, is one of the few sharks to enter estuaries. It feeds on fish, crabs, and shellfish. Young Brown sharks are about two feet long at birth. This is one of the most common large sharks off the coast of New York and New Jersey.

As voracious a feeder as it is, the Tiger shark, *Galeocerdo cuvieri,* has an unfounded reputation as a man-eater. Actually, this shark has a proclivity for meat scraps tossed into the sea from ships, but it will eat almost anything. It uses its wide jaws to consume fish, lobsters, and sea turtles. The Tiger shark is found mainly in the waters off Florida, but during the summer, a few stray further north.

The Atlantic sharpnose shark, *Scoliodon terrae-novae,* lives in shallow water from North Carolina to Florida. It feeds primarily on small fish, shellfish and shrimp. Newborn Sharpnose sharks are about a foot long.

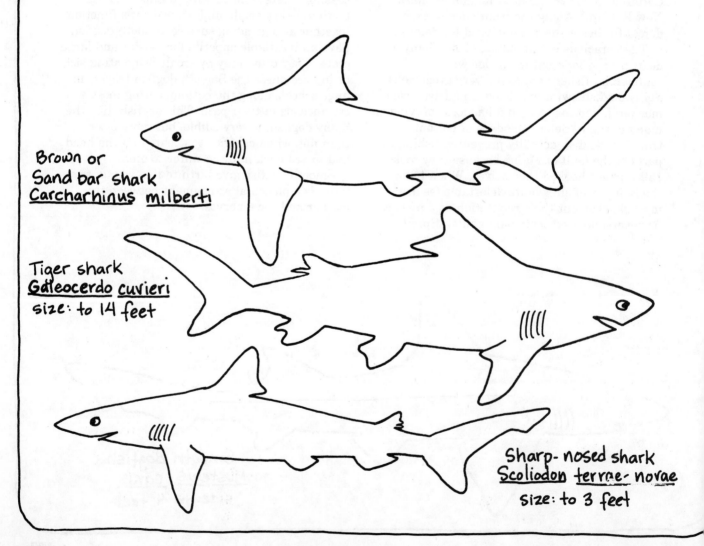

Brown or
Sand bar shark
Carcharhinus milberti

Tiger shark
Galeocerdo cuvieri
size: to 14 feet

Sharp-nosed shark
Scoliodon terrae-novae
size: to 3 feet

Mackerel Sharks

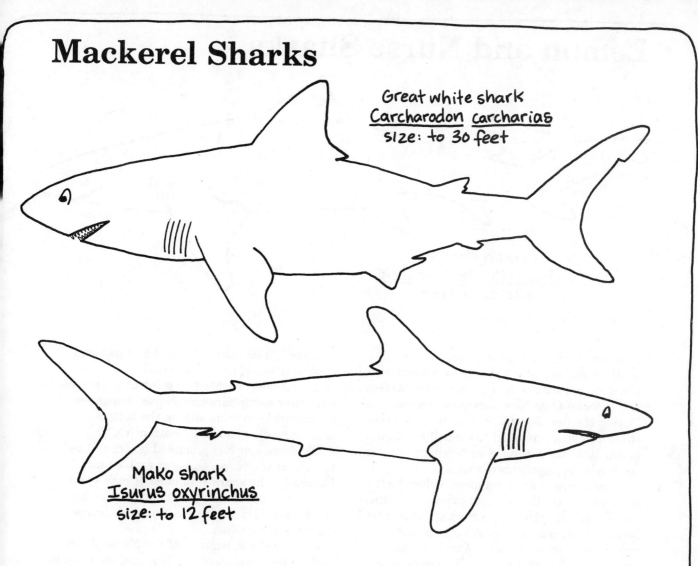

Great white shark
Carcharodon carcharias
size: to 30 feet

Mako shark
Isurus oxyrinchus
size: to 12 feet

Members of the family Lamnidae are known as Mackerel sharks. Mackerel sharks have a crescent-shaped (lunate) tail like that of a mackerel (hence their name) or swordfish. The lower lobe of the tail is only slightly shorter than the upper lobe. Other prominent features of Mackerel sharks are their blade-like teeth and large gill slits.

Undoubtedly the most celebrated cartilaginous fish is the Great white shark, *Carcharadon carcharias,* affectionately known as the Man-eater. The Great white shark is not necessarily white. Although its belly is usually whitish, its back can be very dark, almost black. Behind each pectoral fin is a black blotch.

Great white sharks are typically oceanic, occasionally coming inshore for various unknown reasons. They are found erratically in warm waters around the world and have been found from the Gulf of Maine south to Florida.

Their reputation for devouring people is quite overblown, but it has happened. These sharks are very active, voracious feeders. Small fish frequently fall prey to the Great white shark, but sharks, seals, tuna, and sea turtles have been found nearly whole in the shark's stomach. The teeth of the Great white *are* menacing: they are huge and triangular, with serrated cutting edges. A twenty-one-foot long Great white shark was found to weigh over 7000 pounds, 1000 pounds of which were its liver.

Mako sharks, *Isurus oxrinchus,* live in offshore waters from Maine to Florida. Their teeth are rather cat-like and the front ones flex backwards in towards the mouth. Makos are swift swimmers, thanks to their powerful lunate tail. They are known to leap out of the water when hooked (and sometimes when not hooked). Small schooling fishes such as mackerel and herring are their primary food.

Lemon and Nurse Sharks

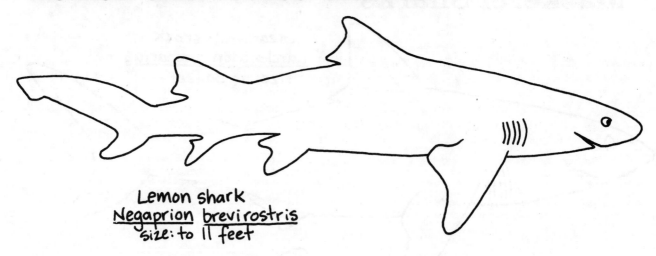

Lemon shark
Negaprion brevirostris
size: to 11 feet

The Lemon shark, *Negaprion brevirostris*, is another member of the family Carcharhinidae, the Requiem sharks, It prefers inshore waters and is found from New Jersey to Florida. During the first few years of its life, the Lemon shark remains restricted to a small area, but by maturity it has expanded its territory to include reefs and deeper offshore waters.

Lemon sharks are born alive. Before birth the embryos obtain food and oxygen from their mother via the placenta. Young specimens feed primarily on crustaceans. Adults eat fish, crustaceans, and even some Turtle grass. Despite the fact that stingray spines become embedded in their jaws, large Lemon sharks devour stingrays. Lemon sharks can "rest" motionless on the bottom, but they burn about 10 percent more energy resting than swimming. It takes a lot of energy for the resting shark to actively pump water over its gills.

Although its mouth is small and its teeth relatively weak, the Nurse shark, *Ginglymostoms cirratum*, has bitten people, but only after being harassed. Nurse sharks are frequently seen lying still on the bottom, looking meek, placid and inattentive. This docile demeanor has prompted ignorant divers to poke at the Nurse shark or grab its tail. Naturally, the shark retaliates. It may swim away or it may clamp onto its tormentor and stubbornly hold on. Unless provoked, Nurse sharks are harmless.

Characteristic features of the Nurse shark include its large upper tail lobe and the barbels hanging below its mouth. Nurse sharks feed on invertebrates on the bottom. Their small teeth hold the prey while their broad jaws crush them. In the summer of 1982, a large Nurse shark was seen off the coast of Ogunquit, Maine, but they are generally found from Cape Cod to Florida.

Nurse shark
Ginglymostoma cirratum
size: to 14 feet

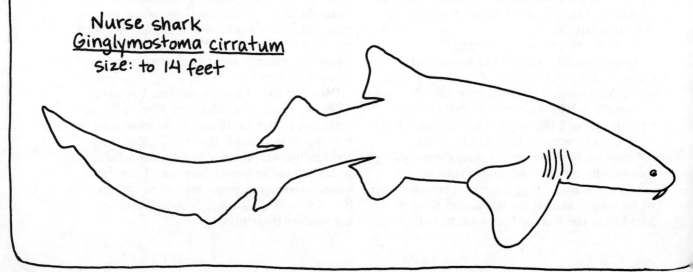

Hammerhead and Bonnethead Sharks

Another shark with an unfounded reputation as a man-eater is the Hammerhead. It's true that on at least one occasion parts of a human being were found in the Hammerhead's stomach, but it would be erroneous to conclude that the shark consumed a live swimmer. Sharks are scavengers and are attracted to dead and dying creatures. Perhaps the consumed person was already dead.

Hammerheads, *Sphyrna zygaena,* and Bonnetheads, *Sphyrna tiburo,* are very similar to Requiem sharks in body shape. The anomalous head sets them apart. Special extensions of the cranium (brain case) around the eye and nose support the "wings" on either side of the head. There are several possible advantages to this unusually shaped head. The broad head may act as a bow rudder, increasing the shark's maneuverability. It may compensate for the poor lift of the shark's relatively small pectoral fins. The distribution of sensory pores all over the wide head may help the shark locate food.

Bonnetheads are found from New England to Florida. They usually live in shallow water, where they feed on crabs, shrimp, cephalopods, and fish. It is said that this shark is somewhat sluggish and aloof. Regardless of its temperament, it is totally harmless to people. Because it reaches sexual maturity at a small size, is easily captured, and lives well in captivity, the Bonnethead is often used in live animal studies.

Hammerheads live from Cape Cod to Florida, occasionally straying up to Maine. In the summer, large schools have been seen off the coast of New York and New Jersey. Hammerheads eat fish (including stingrays) and crustaceans. Both Hammerheads and Bonnetheads readily bite baited hooks. Both give birth to live young.

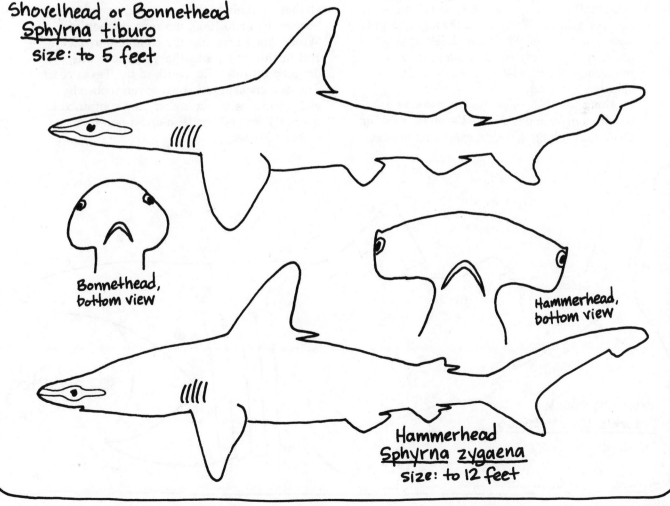

Shovelhead or Bonnethead
Sphyrna tiburo
size: to 5 feet

Bonnethead,
bottom view

Hammerhead,
bottom view

Hammerhead
Sphyrna zygaena
size: to 12 feet

Basking and Whale Sharks

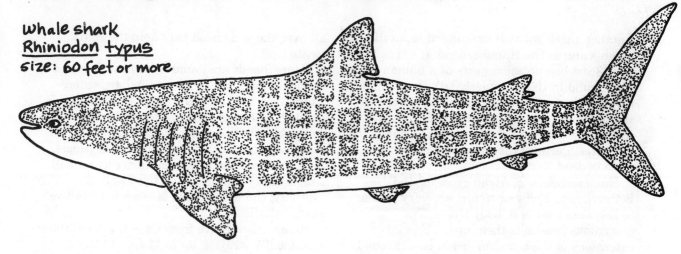

whale shark
Rhiniodon typus
size: 60 feet or more

Whale sharks, *Rhinocodon typus,* are the largest fish in the world; Basking sharks, *Cetorhinus maximus,* are nearly as large. Both feed exclusively on plankton. They don't eat whales, giant squid, ocean liners, or people. The inner margins of the gill arches of both the Whale and Basking sharks are lined with horny, bristle-like gill rakers. To feed, the sharks open their mouth and water pours in. The water rushes over the gills, and planktonic creatures and small fish are caught by gill rakers, which act as strainers. Water exits through the gill slits. The teeth of Whale and Basking sharks are minute (one-sixteenth of an inch) and numerous (thousands), but are no use in chewing or biting.

Along the East coast, Basking sharks are found from Maine to North Carolina. Basking sharks have huge gill slits extending almost

around their neck. Young individuals have a long snout that extends well beyond the mouth. Basking sharks inhabit temperate coastal waters, but are not abundant anywhere. They are usually seen swimming sluggishly or sunning near the surface.

Whale sharks are tropical species that are most abundant in the Caribbean, but they stray as far north as New York. They are rare, solitary creatures. The regular grid of dots and stripes on their back is repeated nowhere else in nature. Prior to 1955, the reproduction of Whale sharks remained a complete mystery. But in that year, a leathery egg case was dredged up 130 miles south of the Texas coast. The case measured twenty-seven inches by sixteen inches by six inches, and contained a perfectly formed fourteen-and-a-half-inch embryo inside.

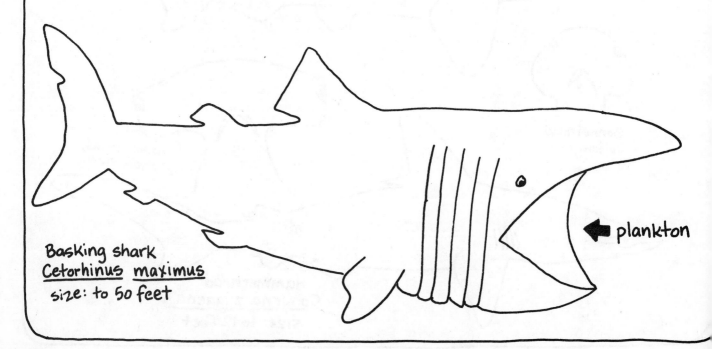

Basking shark
Cetorhinus maximus
size: to 50 feet

◀ plankton

Skates

Skates, sting-rays, manta rays, and eagle rays are all cartilaginous fish. They have flat bodies and wing-like pectoral fins attached to the head.

Skates are common, harmless bottom-dwelling fish. If they breathed through their mouth as most fish do, skates would choke on sand because their mouth is on the underside. Instead, they inhale water through their two spiracles (vestigial gill slits that open and close) near their eyes. Water passes over the gills and out through five pairs of gill slits underneath.

Shellfish, worms, and crabs comprise most of a skate's diet. With their bone-like teeth and well-developed jaws, skates can easily crush crab carapaces and mollusk shells.

Like sharks, male skates have claspers along their pelvic fins to transmit sperm to the female. The fertilized skate eggs develop within a horny capsule that hooks onto seaweed with its tendrils. Empty egg cases, called mermaid's purses, are often found washed ashore.

The Barndoor skate, *Raja laevis*, lives from Maine to South Carolina. Its pectoral fins are sold commercially and are quite tasty. Little skates, *Raja erinacea,* live in shallow water from Maine down to Florida.

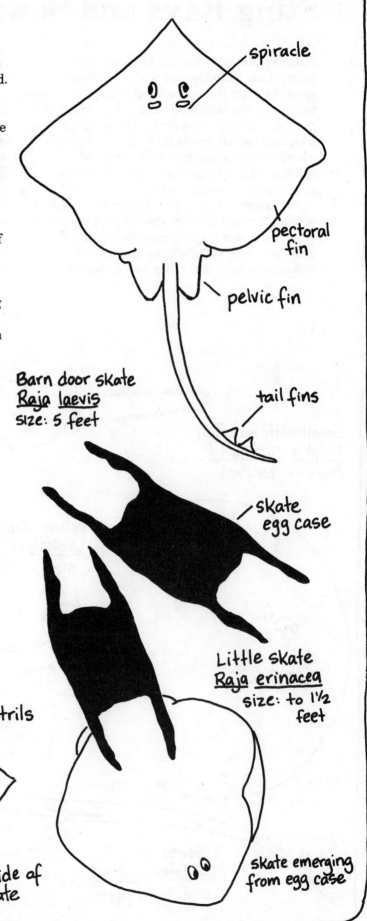

spiracle

pectoral fin

pelvic fin

Barn door skate
Raja laevis
size: 5 feet

tail fins

skate egg case

Little skate
Raja erinacea
size: to 1½ feet

skate emerging from egg case

gill slits

mouth

nostrils

underside of a skate

Sting Rays and Sawfish

Stingrays are very similar to skates, but they have a long whip-like tail with a stiff-edged spine. Usually this spine is held flat against the tail, but if the stingray is stepped on or grabbed, the spine is used defensively. The stingray whips up its tail, erects its spine and impales whatever disturbed it. Venom produced in the spine is released into the wound. To people, this is very painful but not fatal.

Undisturbed stingrays are harmless. When wading through shallow water where stingrays might be present, it is a good idea to shuffle your feet. This will alert the stingray that someone is near, and given half a chance, the ray will swim away. They are not at all aggressive.

Stingrays give birth to live young. The Stingray, *Dasyatis centoura*, lives in coastal water from Cape Cod to Georgia. Yellow spotted stingrays, *Urolophus jamaicensis*, live in Florida waters.

The Smalltooth sawfish, *Pristis pectinata*, is a shark-like ray found in warm waters. With its long, saw-like nose, the sawfish slashes into schools of small fish, which it eats. Sawfish do not have a stinging barb. This species is found in the shallow water of Florida, but strays as far as New York.

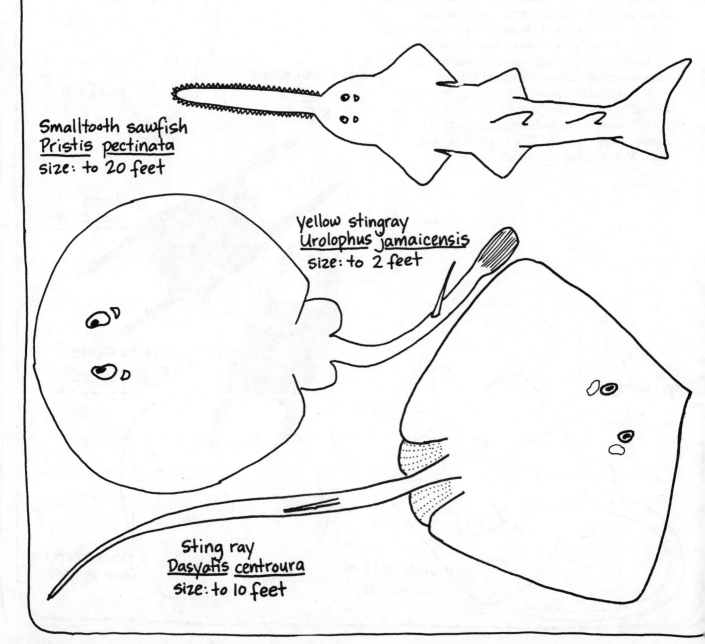

Smalltooth sawfish
Pristis pectinata
size: to 20 feet

Yellow stingray
Urolophus jamaicensis
size: to 2 feet

Sting ray
Dasyatis centroura
size: to 10 feet

Fish True/False Quiz

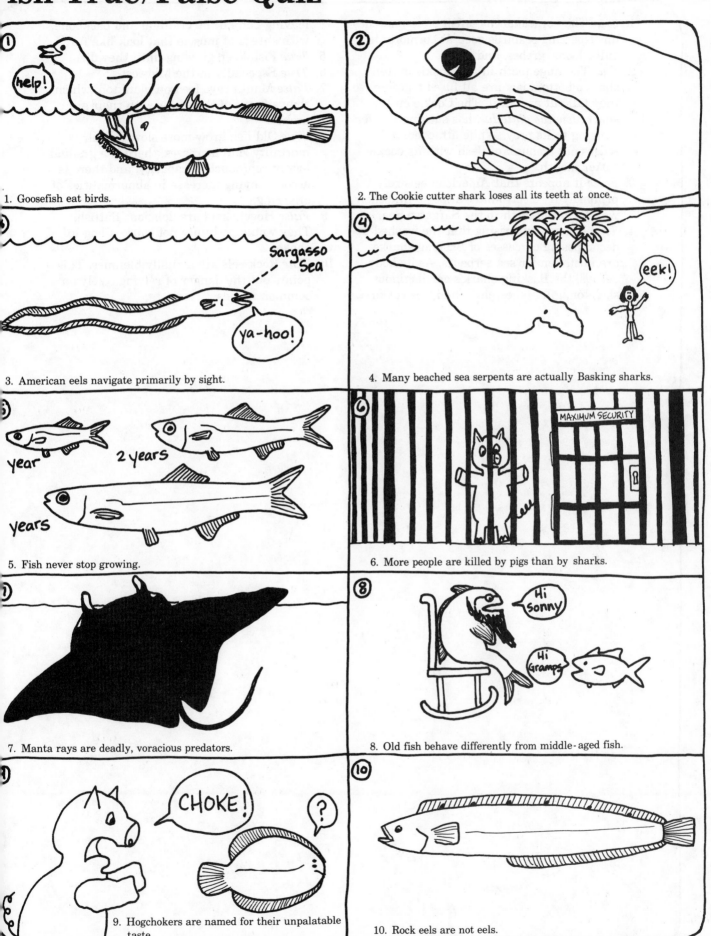

1. Goosefish eat birds.

2. The Cookie cutter shark loses all its teeth at once.

3. American eels navigate primarily by sight.

4. Many beached sea serpents are actually Basking sharks.

5. Fish never stop growing.

6. More people are killed by pigs than by sharks.

7. Manta rays are deadly, voracious predators.

8. Old fish behave differently from middle-aged fish.

9. Hogchokers are named for their unpalatable taste.

10. Rock eels are not eels.

1. *True* The voracious Goosefish often snatches unsuspecting seabirds such as cormorants, gulls, loons, grebes, and auks.
2. *True* The huge teeth in the Cookie cutter shark's bottom jaw are all fused together, so they fall out as a unit. The Cookie cutter shark, *Isistius plutodus,* has sucking lips for clinging to its prey. While attached, it scoops out a chunk of flesh with its cookie cutter mouth.
3. *False* It appears that American eels rely primarily on electro-navigation to find their way back and forth to the Sargasso Sea.
4. *True* It is not surprising that the washed up decomposing carcasses of Basking sharks are mistaken for sea serpents. As its body decays, the Basking shark's cartilaginous skeleton, gill arches, jaws, and fins rot first, leaving behind the cranium and backbone, with shreds of muscle that look like fur.
5. *True* Fish keep growing until they die.
6. *True* Especially in the Midwest.
7. *False* Manta rays feed on plankton, which they sift out of the water with their gill arches.
8. *True* Old fish grow more slowly, their mortality rate increases, there is a gradual loss of reproductive capacity, and there is an accompanying increase in abnormalities of offspring.
9. *False* Hogchokers are delicious flatfish. They would certainly not cause a hog to choke.
10. *True* Rock eels are actually blennies. This points out the danger of relying solely on common names.

5 Marine Reptiles

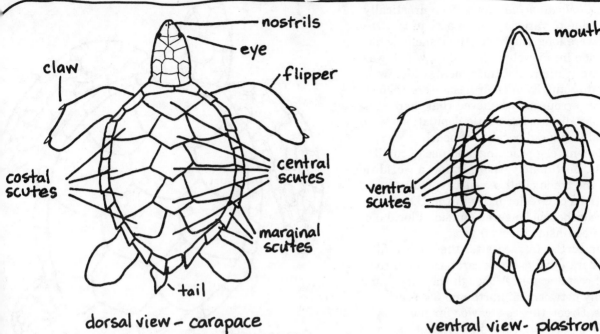

nostrils
eye
flipper
claw
costal scutes
central scutes
marginal scutes
tail

dorsal view - carapace

mouth
ventral scutes

ventral view- plastron

Fourteen percent of the 40,000 known species of vertebrates are reptiles, but very few reptiles live in salt water. Along our coast, the only marine reptiles are the American crocodile and five species of sea turtle. Sea snakes are also marine reptiles, but there are none in the Atlantic.

Reptiles are cold-blooded, air-breathing animals covered with scales. Cold-blooded means the animal derives its body heat from outside sources. By moving to cooler or warmer environments, a reptile is able to alter its body temperature. All reptiles reproduce by internal fertilization. Female sea turtles and crocodiles lay eggs on land, and their young hatch out looking like miniature adults.

The most unique feature of turtles is their lightweight, streamlined shell, which forms an armored enclosure for the vital organs. A turtle's ribs and backbone are firmly attached to the inside of the shell. The upper part of the shell (the carapace) is covered with horny plates called scutes and is connected to the bottom of the shell (the plastron). Sea turtles have heavy necks that cannot be completely pulled into the shell. Their legs are modified into muscular paddle-like flippers for fast (up to 35 mph) and agile swimming.

Sea turtles have lungs and must come to the surface to breathe air, but they have adaptations that allow them to remain submerged for long periods of time. Some sea turtles bring in water through their nostrils and mouth. The lining of the turtle's pharynx acts like a gill and extracts oxygen from the water as it goes down the throat. Other species can take water into their anal opening, where tissues absorb oxygen.

Green and Loggerhead Turtles

As with all sea turtles, populations of Green turtles, *Chelonia mydas*, have been drastically reduced due to various human activities. Green turtles were long sought for their shells, which were made into jewelry, their skin, which was made into leather, and their meat, which was made into soup. Even their nests were robbed, for it was erroneously believed that their eggs were aphrodisiacs. Sea turtle populations were drastically reduced because of beach development, egg poaching, and use of turtle products. Fortunately, there is now a worldwide ban on the trade of all sea turtle products, and populations of Green turtles in the U.S. are listed as either endangered (outside Florida) or threatened (inside Florida).

Green turtles (named after their green fat) have a dark olive to black carapace, with five pairs of costal scutes. Although they are found primarily between 35° north and 35° south latitude, Green turtles stray as far north as Massachusetts. They nest in Florida.

Just prior to nesting time, sea turtles mate at the surface of the water. Female sea turtles can store sperm in their oviduct and fertilize their eggs with it for up to four years. Once females are ready to lay their eggs, they amble up onto a beach and dig a nest in the sand with their flippers. They then lay about a hundred eggs and cover the nest with sand. The hot humid sand incubates the eggs. After one or two months, the baby turtles hatch, usually at night, using a tooth at the end of their beak to break through the shell. Later the egg tooth is lost. As a group, baby turtles push their way out of the sand and try to make it to the sea alive. Less then 1 percent survive to adulthood. Until they are one year old, Green turtles are carnivorous. After one year, they are herbivorous.

The Loggerhead turtle, *Caretta caretta*, has a large log-like head and a red-brown carapace with five pairs of costal scutes. Its blunt beak and powerful jaws are used to crush the crustaceans and mollusks it eats. It also feeds on jellyfish and sponges. Loggerheads nest in the summer on beaches from New Jersey to Texas, and have been seen swimming as far north as Nova Scotia. They are listed as endangered species worldwide.

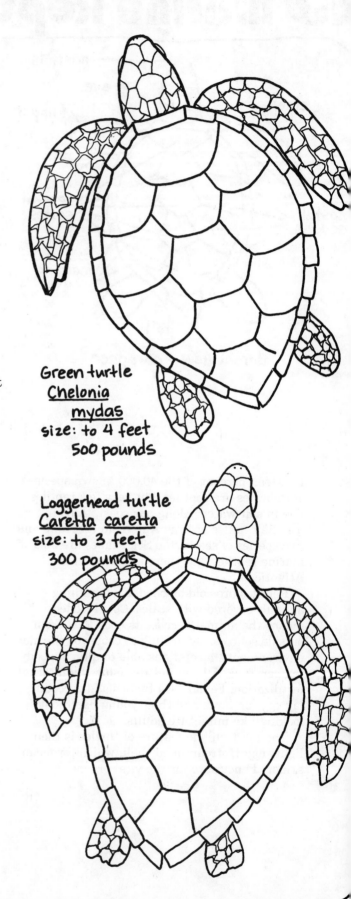

Green turtle
Chelonia
mydas
size: to 4 feet
500 pounds

Loggerhead turtle
Caretta caretta
size: to 3 feet
300 pounds

Hawksbill and Leatherback Turtles

Hawksbill turtles, *Eretmochelys imbricata*, are endangered throughout their range and in the United States. Along our coast they are seen occasionally as far north as Massachusetts, but they definitely prefer warmer water, and are most abundant in the Caribbean. Hawksbills are so named because their beak-like upper jaw overhangs the lower jaw, looking much like a hawk's bill. Their beautiful shell of overlapping scutes was the source of authentic tortoise shell jewelry, for which huge numbers of Hawksbills were killed.

Most sea turtles are nonchalant unless terribly harassed. Yet, despite its small size, the Hawksbill turtle is feisty and aggressive. Hawksbills are slow swimmers and, unlike the Green turtle, which may swim hundreds of miles to its nesting site, they do not migrate. They usually stay around one particular reef or shoal. Hawksbills are omnivorous and feed on algae, grass, fish, and sponges. They also enjoy sea urchins and Portuguese man-of-war!

Leatherbacks, *Dermochelys coriacea*, are large (up to 1500 pounds), torpedo-shaped sea turtles with large flippers. Their carapace is leathery and scaleless, and has five ridges. Unlike all other sea turtles, Leatherbacks have no claws on their flippers. Males have longer tails than females (this is true of other sea turtles also). The shell and bones of Leatherbacks contain large amounts of oil, for which these turtles were killed. Leatherbacks are listed as threatened species under the U.S. Endangered Species Act.

Their streamlined bodies and strong flippers make Leatherbacks strong swimmers. They stay in deep offshore waters and have wandered as far north as Canada. Within the span of a year, a Leatherback may swim back and forth between New England and South America. Many nest in the Caribbean, although rarely in Florida. Leatherbacks feed primarily on jellyfish, but they also eat fish and other invertebrates.

Hawksbill turtle
Eretmochelys imbricata
size: to 3 feet, 160 pounds

Leatherback turtle
Dermachelys
cariacea
size: to 8 feet
900 pounds

American Crocodile

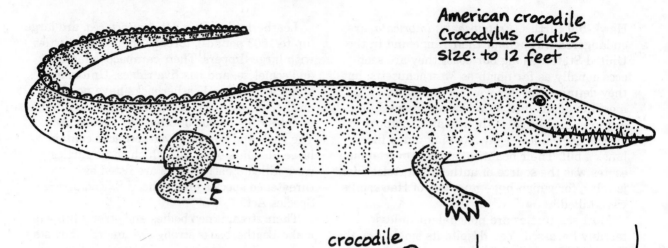

American crocodile
Crocodylus acutus
size: to 12 feet

crocodile

alligator

American alligators and American crocodiles are both heavy-bodied reptiles with flat heads and muscular tails. Three important features help to differentiate between the two. The crocodile has a narrow, tapering snout; alligators have a broad, rounded snout. Both animals have large fourth teeth on each side of the lower jaw, but in the American crocodile, these teeth protrude outside the mouth when the mouth is closed, while in the alligator, they fit into sockets in the upper jaw. Alligators in the U.S. live in fresh to brackish water; crocodiles live in salty water.

About 500 American crocodiles, *Crocodylus acutus,* live in the U.S. today, exclusively in extreme southern Florida and the Keys. After being over-hunted for years for their hide, American crocodiles are now endangered species. Despite this protective designation, the crocodile is still in danger of extinction because of poaching and habitat destruction.

The eyes, nostrils, and ears of crocodiles are located high on their skull and stay above the surface of the water while the rest of the body stays below it. A transparent third eyelid covers the eyes when the crocodiles swim underwater. As crocodiles' teeth become dull or broken, they are shed and immediately replaced. Young crocodiles eat insects, frogs, and spiders. Adults feed on fish, snails, and assorted mammals, such as goats, sheep and cattle.

Crocodiles are fairly aggressive and may bite or swing their tail to defend themselves.

Female crocodiles build nests out of sand or vegetation. Nests can be up to one foot across and two feet high. Eggs are buried about one foot below the surface. Twenty to sixty eggs are laid in the spring, and hatch about three months later. Periodically, the mother crocodile rests her head over the nest. When she senses the young crocodiles are hatching, she digs them out of the nest, picks them up with her jaws, and releases them at the shore. Hatchlings are nine inches long and are preyed upon by lizards, raccoons, wading birds, adult crocodiles, and rogue sharks.

Marine Reptile True/False Quiz

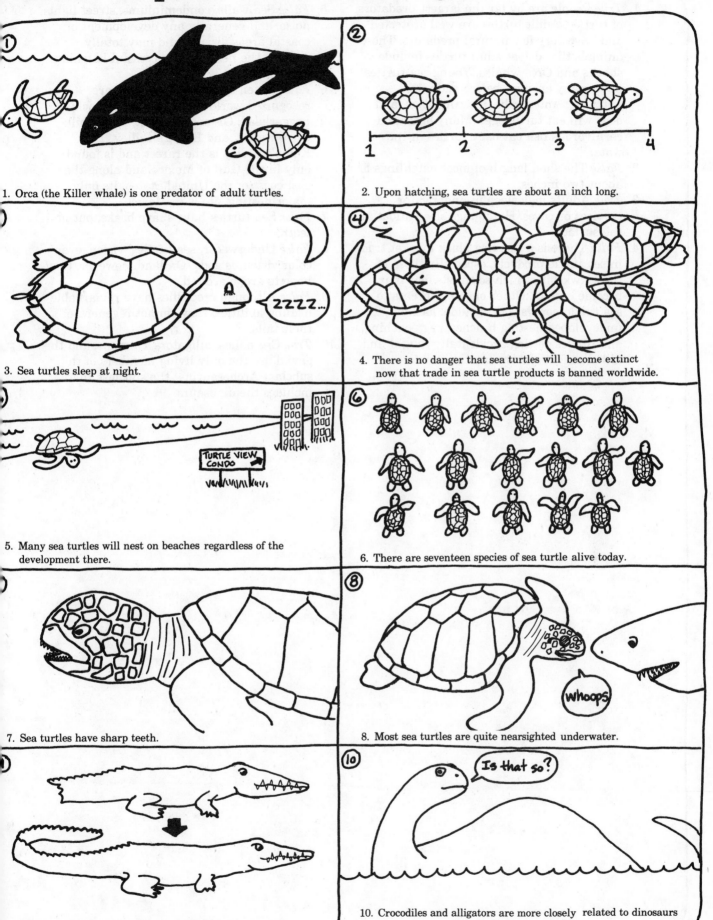

1. Orca (the Killer whale) is one predator of adult turtles.

2. Upon hatching, sea turtles are about an inch long.

3. Sea turtles sleep at night.

4. There is no danger that sea turtles will become extinct now that trade in sea turtle products is banned worldwide.

5. Many sea turtles will nest on beaches regardless of the development there.

6. There are seventeen species of sea turtle alive today.

7. Sea turtles have sharp teeth.

8. Most sea turtles are quite nearsighted underwater.

9. Like most lizards, crocodiles can regenerate their tails.

10. Crocodiles and alligators are more closely related to dinosaurs than are any other animals alive today.

1. *True* People are by far the largest predators of turtles. Adult turtles are well-protected and have very few natural predators. Those animals that do eat adult turtles include sharks and Orca whales. Young sea turtles have many enemies before and after hatching. Ants, crabs, raccoons, pigs, and dogs will eat turtle eggs. Many birds and carnivorous fish eat baby turtles in the water.
2. *False* The shell length of most hatchlings is about two inches.
3. *True* Adults may sleep in underwater crevices or caves. Hatchlings sleep afloat at the surface.
4. *False* There is a thriving black market trade in sea turtle products, especially those of the Hawksbill, which scientists fear may become extinct. Other ongoing threats to sea turtle populations include: lack of undeveloped nesting beaches; egg poaching, and turtles getting accidentally caught and killed in fish and shrimp trawl nets.
5. *False* Sea walls, condominiums, street lights, noise, and generally any development of coastal areas disrupt and may totally prevent the nesting or successful hatching of sea turtles.
6. *False* Seven species of sea turtle are recognized worldwide: the Leatherback; Loggerhead; Green; Flatback; Hawksbill; Olive Ridley; and Kemp's Ridley. The Kemp's Ridley is the rarest and is found only in the Gulf of Mexico and along the east coast of the United States. Its only major nesting site is in Mexico.
7. *False* Sea turtles have sharp beaks, but no teeth.
8. *False* Underwater, sea turtles' vision, even color vision, is good. On land, however, they have to grope around.
9. *False* Although crocodiles have remarkable healing abilities, they cannot regenerate their tails.
10. *True* Crocodiles, alligators, caimans, and the gavial are the only living members of the subclass Archosauria—the subclass that includes the dinosaurs.

6 Birds of Sea and Shore

Birds are warm-blooded vertebrates. Fossil evidence indicates that birds have evolved from reptiles; their anatomy and embryology are similar. Most birds can fly, and all have feathers, which form a durable, light-weight, insulating covering. Generally, a bird's plumage is waterproof. While preening, birds rub their beaks, feet, and feathers with oil from a gland near their tail. The oil helps waterproof the feathers and keeps the beak and feet lubricated. Nearly all birds completely molt once a year, usually after nesting.

Skeletons of vertebrates, birds included, protect the internal organs and provide anchorage for the muscles. Bird bones are light-weight, yet strong, and air-filled. Many are fused together, resulting in a rigid skeleton necessary for flight. Birds have light beaks and jaws, but no teeth.

Although birds have a minimal sense of taste and smell, their hearing and sight (even color vision) are acute. A nictitating membrane, acting as a third eyelid, cleanses and protects the eyeball. Birds inhale air through the nares (nostril openings) on their beak. Excess salt that seashore birds take in through drinking and eating accumulates in glands near their nostrils and oozes out of the nares. Aquatic birds generally do not have salivary glands. Birds have a high body temperature (100°–112° F), high metabolism, and a strong four-chambered heart that maintains a rapid heart rate.

After male and female birds mate, the eggs in the female pass to the oviduct, where they are fertilized and then covered with albumen ("white") and a calcium shell. Females lay fertilized eggs, usually one a day, and incubate them (with or without the father's help) until they hatch.

Most birds in the northern hemisphere migrate south (of their breeding range) in the winter and return north to breed in the summertime.

The Atlantic puffins (*Fratercula arctica*) breed from Greenland to the Maine coast, but in the winter they may go as far south as Massachusetts. These birds are chunky and short-necked, and have a triangular bill that is

Atlantic puffin
Fratercula arctica
size: to 1 foot

covered with a colorful sheath during breeding season. Populations of puffins have declined near populated areas. Their tame, curious nature made them easy prey for hunters, who shot them for food. Puffins nest in colonies on offshore islands. They lay eggs in burrows, but rats, cats, and dogs eat many of these eggs. Puffins have disappeared from islands where these predators are present.

Puffins swim underwater with their wings, pursuing fish, mollusks, and other sea creatures for food. On land they walk upright easily. Although they don't look it, puffins are perfectly capable of flying.

Shorebirds

Many species of birds wade into shallow water to find their prey, but do not swim. At first glance these birds may look similar, especially the sandpipers. Even at second glance, some shorebirds are hard to tell apart. Shorebirds are usually countershaded—dark above, where they receive the most light, and whitish below, where they receive the most shadow. This coloration makes them hard to see on the beach and hard to tell apart by color alone. Upon closer observation, however, each species can be identified by its distinctive bill, legs, and behavior. The species on this page are only a few of the hundred or so birds seen on beaches, tidal flats, rocky shores, jetties, and marshes.

The Least sandpiper, *Calidris minutilla*, is the smallest of the seemingly billions of species of sparrow-sized shorebirds. (For those who have neither the patience nor the inclination to differentiate between the plethora of small shorebirds, they are all referred to as "peeps.") It winters all along the Atlantic coast and breeds in Alaska and Canada. All sandpipers are wading birds with slender bills used to probe into shallow water (or mud) for crustaceans, worms and mollusks.

A noisy but wary bird that may wade out past its waist is the Greater yellowlegs, *Tringa melanoleuca*. The yellowlegs runs around preying on fish and crustaceans, seldom probing into the mud. It winters all along the coast and breeds in Alaska and Canada.

Ruddy turnstones, *Arenaria interpres*, breed in the arctic and winter all along the East coast. They are short, squat and pugnacious, and eat whatever they find under the rocks and seaweed they overturn.

The American oystercatcher, *Haematopus palliatus*, was nearly eradicated from the U.S. coast by overzealous hunters. Now, with protection, it has become more numerous and is found from Cape Cod south to Florida. This chicken-sized bird has a long, flat, chisel-like red bill. It opens bivalves by sticking its bill into a bivalve that is partly opened and cutting the adductor muscle before it can close.

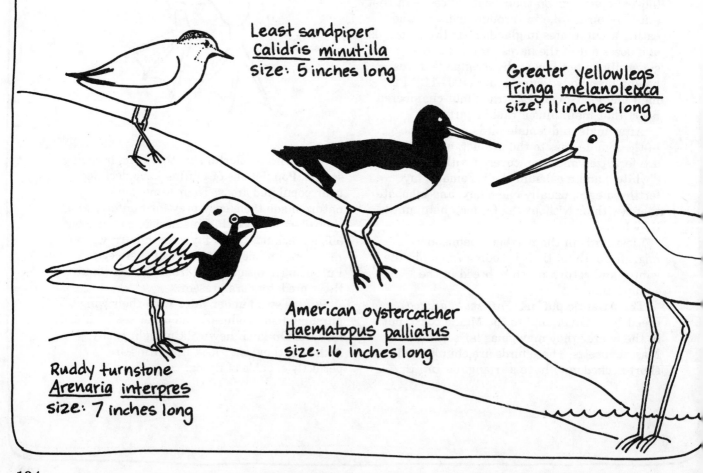

Least sandpiper
Calidris minutilla
size: 5 inches long

Greater yellowlegs
Tringa melanoleuca
size: 11 inches long

American oystercatcher
Haematopus palliatus
size: 16 inches long

Ruddy turnstone
Arenaria interpres
size: 7 inches long

Egrets, Herons, and Ibis

Herons, egrets, and ibises are wading birds with extremely long legs. They have long necks and bills used to make sudden thrusts in the water or to probe in the mud. Widespread toes keep the birds from sinking in mud. In flight, herons' and egrets' necks are kept folded like an "S"; the ibis flies with its neck outstretched.

The Snowy egret, *Egretta thula,* breeds and lives from the northern U.S. south to Florida. During the nineteenth and twentieth centuries, Snowy egrets were almost hunted to extinction for their fine recurved back plumes (breeding plumage), which were valued as hat ornaments. With protection, Snowy egrets are once again common birds of the shore and marsh. They have a slender black bill, black legs and bright yellow feet. Snowy egrets are agile feeders and rush around in the water stirring up food (fish, crustaceans, etc.) with their feet.

Reddish egrets, *Dichromanassa rufescens,* were also killed for their stylish plumes. They are still not common, partly because their nesting is easily disturbed by intruders and curious bird watchers. There are two color phases of the Reddish egret. In Florida, they are grey with a rusty head and neck. The Texas populations are white with blue legs. Both have rather shaggy necks. Reddish egrets dance and lurch about in the water to scare up prey.

The Great blue heron, *Ardea herodias,* may grow to four feet tall. This beautiful stately bird lives and breeds from Southern Canada to Florida. Most migrate south in the fall. Great blue herons have a varied diet and eat mice and snakes in addition to the usual fish, crustaceans, and worms.

The White ibis, *Eudocimus albus,* is very heron-like except for its decurved bill. Like the herons and egrets, it nests with other shore birds in large colonies in low trees or shrubs near the water. The White ibis lives around marshes and mangrove swamps in the southeast U.S.

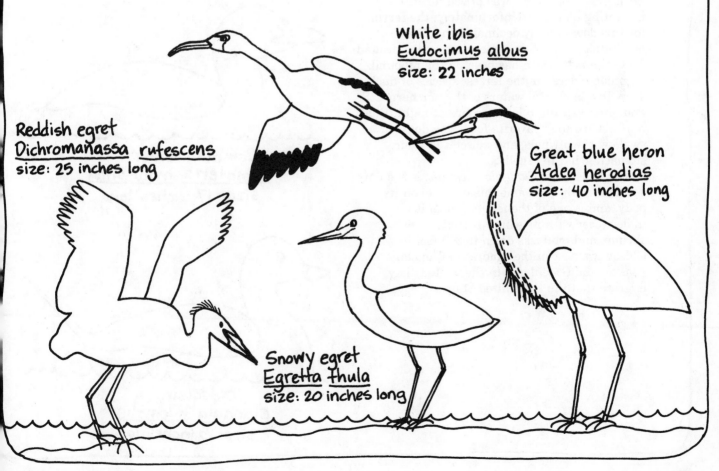

White ibis
Eudocimus albus
size: 22 inches

Reddish egret
Dichromanassa rufescens
size: 25 inches long

Great blue heron
Ardea herodias
size: 40 inches long

Snowy egret
Egretta thula
size: 20 inches long

Sea Ducks

Ducks are not usually associated with the sea, but many species are common in bays, estuaries, and open shores. All the ducks shown here are divers: instead of feeding at the surface (dabbling ducks), they dive underwater to pursue prey.

The Red-breasted merganser, *Mergus serrator*, has a spike-like bill with saw-tooth margins. It lives mainly on fish, which it chases underwater, but it also eats crustaceans. This merganser breeds in the northern U.S. and Canada, and spends the winter in salt water from Maine to Florida. The other two species of merganser, the Common and the Hooded, are not found in salt water.

Most ducks propel themselves underwater with their feet. The White-winged scotor, *Melanitta deglandi,* also uses its wings. This is a heavy-looking black duck with white wing patches that may not be visible until the duck flies up. It breeds in Alaska and Canada, and spends its winters along the U.S. coast. Mollusks are its primary food, and the scotor dives fifteen to twenty-five feet down to collect them.

The down of the female Common eider, *Somateria mollissima,* was prized for its insulating qualities. Unfortunately, the fervor for this down nearly decimated the Eider population. Since this persecution ended around 1900, the population has been on the rebound. Common eiders are the largest species of duck. They live in coastal waters of the northern U.S. and often raft up in large groups at night. Blue mussels are their favorite food. The ducks swallow them whole, and muscles in their stomach grind the shells.

The Oldsquaw, *Clangula hyemalis,* is the only sea duck with large amounts of white on its body, and is one of the few to propel itself underwater with its wings. It feeds on fish, shrimp, and mollusks down to 200 feet. Oldsquaws nest in the tundras of Canada, Alaska, and Greenland. In the winter, they migrate south to the United States.

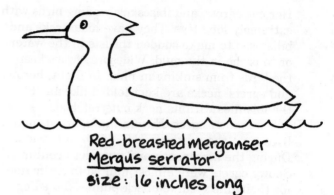

Red-breasted merganser
Mergus serrator
size: 16 inches long

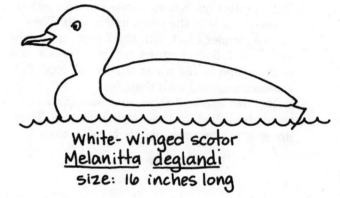

White-winged scotor
Melanitta deglandi
size: 16 inches long

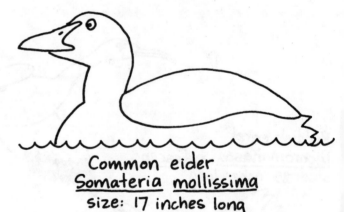

Common eider
Somateria mollissima
size: 17 inches long

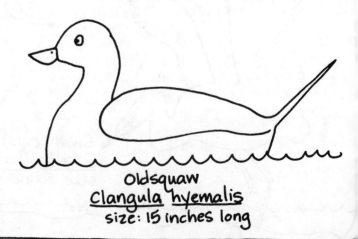

Oldsquaw
Clangula hyemalis
size: 15 inches long

Gulls, Terns, and Skimmers

Common tern
Sterna hirundo
size: 14 inches long

Black skimmer
Rhynchops niger
size: 17 inches long

Laughing gull
Larus atricilla
size: 13 inches long

Herring gull
Larus argentatus
size: 20 inches long

Gulls are long-winged birds with slightly hooked bills. They swim, but seldom dive, and are omnivorous. The Herring gull, *Larus argentatus,* is the common "seagull" seen along the shore south to the Carolinas and inland near large bodies of fresh water. These gulls are voracious scavengers with a proclivity toward garbage. As garbage dumps increase, so does the Herring gull population. The expanding population is detrimental to other marine birds because the Herring gull takes over their nesting sites and eats their eggs. Herring gulls also feed on crabs and shellfish. By dropping crabs and mussels on the pavement while flying, the gulls break their shells. At night the gulls congregate in large roosts in protected waters, on beaches, or on islands.

Laughing gulls, *Larus atricilla,* aren't such avid scavengers. Crabs and fish are their preferred food. These birds have a laugh–like call; *Ha ha ha ha ha,* and are common during the summer along the Atlantic and Gulf coasts. In the summer, their head is light; in the winter it is dark.

Terns are more steamlined than gulls, and usually have forked tails. Their bill is kept pointed down toward the water during flight. Unlike the gulls, terns hover in the air, then dive into the water for fish. Most species of tern have a black cap in the winter and a white cap in the summer. Common terns, *Sterna hirunda,* are pigeon-sized birds with deeply forked tails. Colonies of Common terns nest on islands or isolated peninsulas. Like other terns, this species is very sensitive to human disturbances during nesting. Nesting sites must be protected to maintain the population.

Black skimmers, *Rynchops niger,* are crow-sized relatives of gulls and terns. They have a large, red knife-like bill, the lower half (lower mandible) of which is much longer than the upper half (upper mandible). Skimmers fly low over the water and dip their lower mandible in to catch fish and crustaceans. Black skimmers breed from Cape Cod to Florida; they spend their winters on the southeast coast.

Cormorants

Watching cormorants can provide hours of entertainment and education. The species most likely to be seen from Maine to Florida is the Double-crested cormorant, *Phalacrocorax auritus,* a black goose-like bird with an orange throat pouch and a barely noticeable double crest on its head. In the winter, the Double-crested cormorant heads south (but stays as far north as Long Island), and the Great cormorant, *Phalacrocorax carbo,* comes down from Canada into New England. The Great cormorant is larger and has a white patch under its throat.

Cormorants fly with their necks out-stretched, and occasionally glide when flying. Their takeoff from the water is less than graceful. Usually they hit the water with their wings several times before gaining any altitude. While swimming, cormorants tilt their bill upward in a dignified manner.

Cormorants eat fish. To capture their prey, cormorants dive underwater from a swimming position, not from the air. Using their webbed feet (and occasionally their wings) for propulsion and their tail as a rudder, cormorants swim underwater and nab fish. Cormorants have very flexible throats, and by tilting their head up they can swallow surprisingly large fish. Their bones are rather heavy; this helps the bird submerge. When cormorants emerge from the water after eating, they perch upright and extend their wings to dry. Their plumage is not waterproof.

Cormorants produce an incredible amount of guano. They rapidly convert their fish dinners into nutrient-rich (nitrogen and phosphorus) droppings. Waters around nesting colonies of cormorants have luxuriant algal growth that feeds large populations of fish and invertebrates. Along the northeast coast, Double-crested cromorants nest on cliffs or rocky islands, often near nests of gulls and eiders. In the south, this cormorant nests in trees near the water, adjacent to nesting herons and pelicans. One tree may contain over thirty cormorant nests.

Double-crested cormorant
Phalacrocorax auritus
Size: 27 inches long

Pelicans

Pelicans are large, robust birds that feed on fish and crustaceans. Hanging under their long, flat bill is a voluminous, fleshy throat pouch, which is inflated only when the pelican is underwater snatching prey. Contrary to popular opinion, the pelican's pouch is not used to carry or store fish. It serves as a scoop to separate the fish from the water. When deflated, the pouch is flat and not noticeable.

Populations of Brown pelicans, *Pelacanus occidentalis,* were nearly obliterated due to insecticides (such as DDT), which accumulate in the fish the pelicans eat. The insecticide caused the pelican's eggs to be thin-shelled or broken, and young pelicans died before hatching. Fortunately, the Brown pelican population is now recovering.

Brown pelicans have a wingspan of about six and a half feet. They are strictly coastal and are found from the North Carolina to Florida and the Gulf of Mexico. When searching for fish, pelicans cruise close to the water, gliding for long distances with only an occasional flap of their wings. Once they spot food, the Brown pelicans ascend into the air and drop straight down head first into the water, producing a large splash. Usually they feed on fish that school near the surface, such as mullet, herring, and menhaden. Frigate birds, terns, and gulls constantly harass the pelicans and try to steal their food. Pelicans nest in trees or on the ground of small islands in shallow bay areas. Any human disturbance of the nesting site during breeding is very detrimental to successful hatching.

White pelicans, *Pelacanus erythrorhynchus,* have an incredible nine-foot wingspan. They breed along the West coast and in the central United States near lakes, marshes, and beaches. In the winter they migrate to the Everglades and Florida Keys. Unlike the Brown pelicans, the White pelican scoops up fish while swimming rather then by dive bombing in from the sky.

White pelican
Pelacanus erythrorhynchos
size: 50 inches long
110 inch wingspan!

Brown pelican
Pelacanus occidentalis
size: 40 inches long
90 inch wingspan

More Birds

The Northern gannet, *Morus bassanus,* is a sleek, goose-sized bird that breeds in colonies off the coast of Canada. It lays one egg in a nest of seaweed high on a cliff of an island. In the winter, gannets are found in coastal waters as far south as Florida. Gannets rarely, if ever, come in close to shore (except to breed), but they can be seen in the distance flying or plunging head first for fifty feet or more down into the water. Special air sacs under their breast serve to cushion the blow of their dive. Although their aerial dive propels them quite a distance underwater, the gannet may also use its wings while swimming after prey. Gannets eat all types of fish, especially those that congregate in large schools near the surface.

The Magnificent frigatebird, *Frageta magnificens,* has such short legs and weak feet that it can barely walk, and cannot take off except from a high rock or tree. It never lands on water or land. In contrast to its puny legs, frigatebirds have the largest wingspan in proportion to their weight of any bird. With outstretched wings, they soar and then swoop down to catch fish that leap out of the water. Frigatebirds often harass seabirds, causing them to drop their food, which the frigatebird deftly snatches before it hits the water. Frigatebirds live in shallow bays and mangrove swamps in Florida and the Gulf of Mexico. They are also called Man-of war birds.

Many birds, such as Leach's storm petrel, *Oceanodroma leucorhoa,* are truly pelagic, and are rarely within sight of land. This bird is small and swallow-like. Some pelagic birds follow boats; this one seldom does. Leach's storm petrels are rarely even seen because they fish far out at sea and come to their breeding grounds after dark. In spring they breed on rocky islands and deserted coasts from Canada to Massachusetts, but they spend the winter on the open ocean. They settle on the water to feed and pluck out shrimp and small crustaceans as they go by.

Gannet
Morus bassanus
Size: 30 inches long
70 inch wingspan

Magnificent frigatebird
Fregata magnificens
Size: 35 inches long
90 inch wingspan

size: 8 inches long
20 inch wingspan

Leach's storm petrel
oceanodroma leucorhoa

Bird True/False Quiz

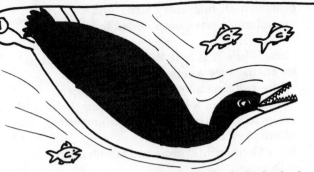

1. Prehistoric birds often dove to get fish, as do the birds of today.

2. All baby birds are born without feathers.

3. Hens are one of the few birds that lay unfertilized eggs.

4. Few birds cross the equator during migration.

5. All native birds in this country are protected by law.

6. Birds do not help each other find food.

7. Cormorants defecate a lot, but never in their own nest.

8. Kingfishers are not found near salt water.

9. Osprey plunge feet first into the water for fish.

10. Baby birds have a special "tooth" they use to open their shell.

1. *True* Fossils found in Kansas (inland seas inundated the land and left deposits when the water receded) include diving birds, such as this five-foot-long comorant-type bird shown in the first panel of page 201. At this stage in their evolution, some birds had teeth.

2. *False* A few birds are born naked, but most hatch with a covering of down. Some birds, such as robins, hatch in an immature stage and have sparsely distributed rows of down. Other birds are completely covered with down and can usually walk soon after hatching.

3. *True* The chicken eggs we buy are unfertilized. Very few birds lay unfertilized eggs.

4. *False* Many birds cross the equator. The equator is not a barrier to migration. Birds use river valleys, mountain chains, and coastlines as lanes for travel. Birds fly between North and South America along special routes and preferred lanes called *corridors*. Many corridors contribute to a major flyway. Arctic terns, Bobolinks, and Golden plovers are just a few of the birds that migrate over the equator.

5. *True* The Migratory Bird Treaty Act of 1918 protected all migratory birds, and amendments have been added to include all native birds.

6. *False* While it is true that most birds fend for themselves where food is concerned, White pelicans may get together to herd fish. They form a long line, beating their wings to drive the fish into shallow water, where they are caught.

7. *False* Cormorants' nests are often covered with guano.

8. *False* Kingfishers fish in salt marsh areas as well as fresh water.

9. *True* The sole of the Osprey's feet are spiny, allowing the Osprey to firmly grip its slippery victim. An Osprey hovers over the water until a fish happens by. It then grabs the fish with its toes (talons).

10. *True* Chicks start to crack their shell by struggling within it. They break out using their "egg tooth," a horny protuberence on the tip of the upper part of the bill. The egg tooth disappears soon after hatching.

7 Marine Mammals

About 350 million years ago, some amphibious creatures left the sea for the land. Over millions of years of gradual adaptations, one group of these creatures evolved into mammals. Some mammals returned to the sea to live; they are the marine mammals of today: the whales, seals, and manatees. Mammals are warm-blooded vertebrate animals that have hair and breathe air. Young mammals develop within their mother (except for the platypus, which lays eggs) and after birth, the mother cares for and feeds them. Mammals have mammary glands that produce milk.

Whales (the Cetaceans) have adaptations enabling them to survive cold temperatures, to dive to great depths, and generally, to make survival easier in the water. They are streamlined (the male's penis and female's mammary glands are tucked inside the body) and swim using up and down movements of their horizontal tail flukes for propulsion. Their paddle-like flippers are used for steering and stabilization. Layers of blubber under the skin insulate the whale, and store energy. The whale's nose (its blowhole) is located on top of its head so it is exposed as soon as the whale surfaces for air. After it exhales old air and takes a breath, the whale pinches its nostrils closed and dives down, coming up for air again two to ten minutes later. While diving to great depths, whales conserve oxygen by cutting off oxygen flow to non-essential areas and slowing down their bodily processes.

The two major types of whales are the toothed and the baleen. Toothed whales feed on large prey; baleen whales strain planktonic food out of the water.

By beaming sound waves through the water, whales can navigate through a sonar-like process called echolocation. They are able to make noise and communicate by forcing air through their closed nasal passages.

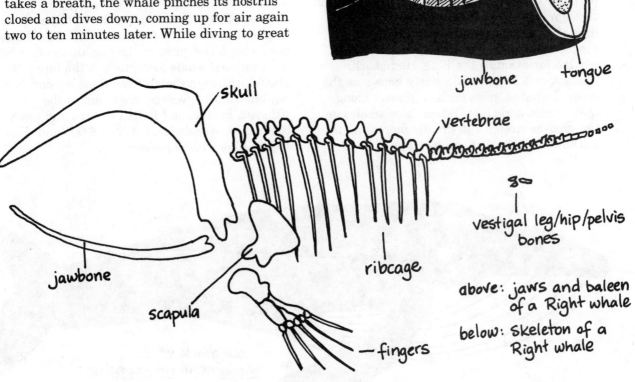

above: jaws and baleen of a Right whale

below: skeleton of a Right whale

Baleen Whales

It has been theorized that during the course of evolution, some whales increased to such a size that teeth became inadequate to catch enough food to sustain the whale's massive body. In all species of great whale except the Sperm whale, teeth were gradually replaced by baleen. Baleen is a series of fringed flexible keratin plates that hang down from the whale's upper jaw. Baleen whales belong to the suborder of Cetaceans called Mysticeti. The term Mysticeti comes from the Greek word for moustache, since the baleen is vaguely reminiscent of a hairy upperlip.

To feed, a baleen whale opens its jaws, and water, teeming with krill and/or other plankton, pours into its mouth. Then the whale partially closes its mouth, forcing the water out through its baleen. Food is trapped on the inside of the baleen and is wiped off with the tongue and swallowed whole. Some baleen whales, for example the Blue, Humpback, Finback and Minke (collectively known as the rorqual whales), have folds or grooves along their throat and chest. When these whales gulp, the grooves stretch out, greatly increasing the amount of water that can be trapped.

Blue whales, *Balaenoptera musculus*, can reach a length of 100 feet, but most of the large ones were killed prior to the beginning of international whale protection in the late 1960's. Although their seasonal movements are not known, Blue whales are found in the Atlantic, Pacific, and Indian Oceans and near both poles. Worldwide, between 6,000 and 10,000 Blue whales remain.

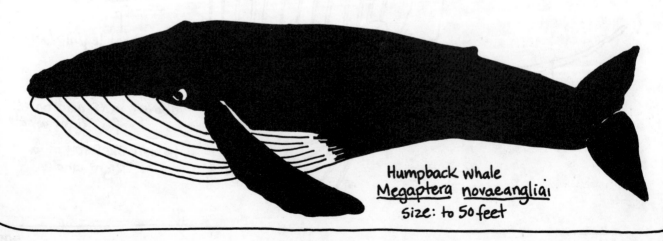

Humpback whale
Megaptera novaeangliai
size: to 50 feet

More Baleen Whales

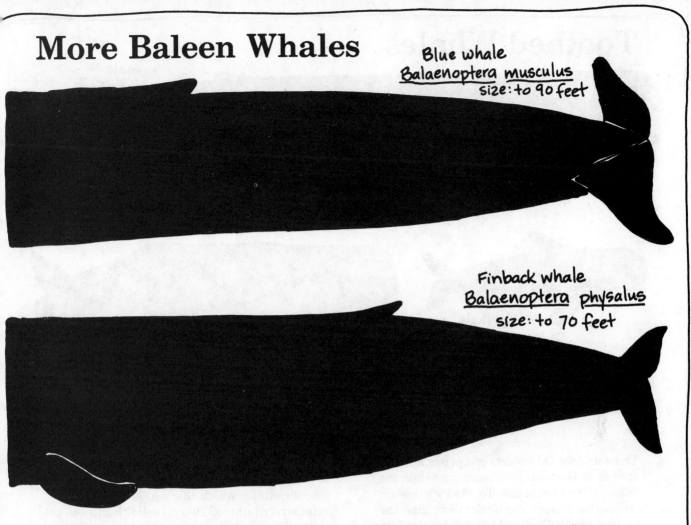

Blue whale
Balaenoptera musculus
size: to 90 feet

Finback whale
Balaenoptera physalus
size: to 70 feet

Finback whales, *Balaenoptera physalus*, live in all oceans, but prefer temperate water, where food is abundant. Humpback whales, *Megaptera noveangliae*, are found worldwide. In the summer, they feed in the productive waters of temperate areas. In the winter, they move to tropical areas to breed and calve.

The 3,000 Right whales, *Eubalaena glacialis*, alive today are sparsely distributed around the world. Right whales were the right whales to kill because they are slow swimmers and their carcasses float.

In the 1860's, steamships and the first practical harpoon gun became available. Then the fast swimming rorquals were hunted and killed by the thousands. Carcasses of rorquals sink, but they were pumped up with air so that they would float. In the heydey of factory whaling ships after World War II, 25,000 Finback whales were captured *each year*. Today there are probably only 100,000 Finback whales left in the world.

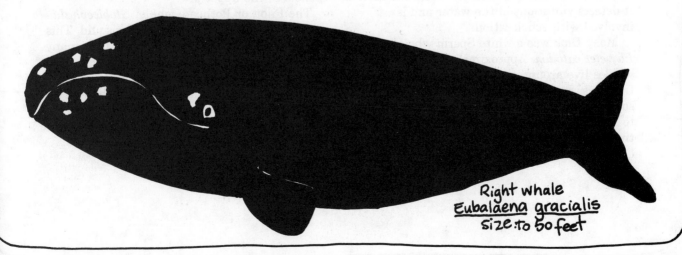

Right whale
Eubalaena gracialis
size: to 50 feet

Toothed Whales

male:
to 50 feet

Sperm whale
Physeter catodon

Pilot whale
Globicephala melaena
size: to 20 feet

female:
to 35 feet

The suborder Odontoceti comprises all sixty-five species of the toothed whales, including the dolphins and porpoises. Toothed whales are smaller and much more prevalent than the baleen whales. Toothed whales are predators, and use their teeth to capture their prey, which they swallow whole. Fish and squid are their primary food.

All whales have a double nasal opening in their skull, but only baleen whales have two *external* nasal openings (thus a double blowhole); toothed whales have one external opening (one blowhole). Another unique feature of the toothed whales is the accumulation of waxy tissue between the top of their head and their upper jaw. When abundant, as in dolphins, this mass of tissue is called the melon. It detects vibrations in the water and is involved with echolocation.

Moby Dick was a white Sperm whale, *Physeter catodon*. Approximately 600,000 Sperm whales live around the world in temperate, ice-free seas. Sperm whales have a well organized social structure and travel in groups of twenty to forty. Several groups regularly migrate along the Atlantic coast between Canada and the Caribbean.

Sperm whales have a massive head (one third of their body length), filled with a wax-like oil, spermaceti, for which this whale was extensively hunted. This oil solidifies in the cold water Sperm whales encounter during their deep dives, and helps reduce the whale's buoyancy. Hunters coveted the spermaceti for its use as long-lasting lamp oil and candle wax. The blowhole of the Sperm whale is on the left side of its forehead, so the whale spout shoots off at a 45° angle.

Although they prefer giant squid, Sperm whales will eat octopi and fish. They are bottom feeders and plow along the bottom (at depths to 2,000 feet) with their lower jaw. About sixty large conical teeth in the lower jaw fit into sockets in the upper jaw.

The Pilot, or Pothead, whale, *Globicephala melaena*, also has a prediliction for squid. This whale has a very bulbous forehead. Like many other whales, the Pilot whale spends the summer feeding in plankton-rich waters and heads to warmer, offshore water in the winter to breed. Occasionally, groups of Pilot whales will become stranded on the shore, perhaps due to an inner ear parasite that impedes their navigation.

Dolphins and Porpoises

Dolphins and porpoises are members of the suborder Odontoceti—the toothed whales. Over the years, the terms dolphin and porpoise have become interchangeable. Most commonly, people refer to those species with a long beak and conical teeth as dolphins. Those species without a beak and with spatulate teeth are called porpoises. Researchers frequently refer to both types of cetaceans as porpoises, thus avoiding any confusion with the Dolphin fish, a popular game fish in Florida.

The Harbor porpoise, *Phocoena phocoena*, is the most common whale in the Gulf of Maine, but it is found as far south as New Jersey. Actually, this porpoise does have a beak, but the melon (deposit of waxy tissue) is located on top of the beak beside the forehead, creating a continuous smooth profile all the way to the mouth. Squid and fish are the preferred foods of the Harbor porpoise.

There are many species of dolphin, but the Bottlenose dolphin, *Tursiops truncatus*, is probably the one familiar to most people. This species is the subject of much study and is a favorite performer at aquariums. All cetaceans are very aware, astute, and intelligent.

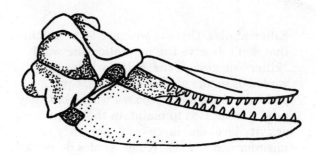

skull of Bottlenose dolphin

Bottlenose dolphins are especially adept at imitating, memorizing, and solving problems. Experiments with their echolocation ability indicate that a blindfolded dolphin can easily maneuver through a maze. Bottlenose dolphins live in tropical and temperate seas. Often they swim and gambol alongside boats.

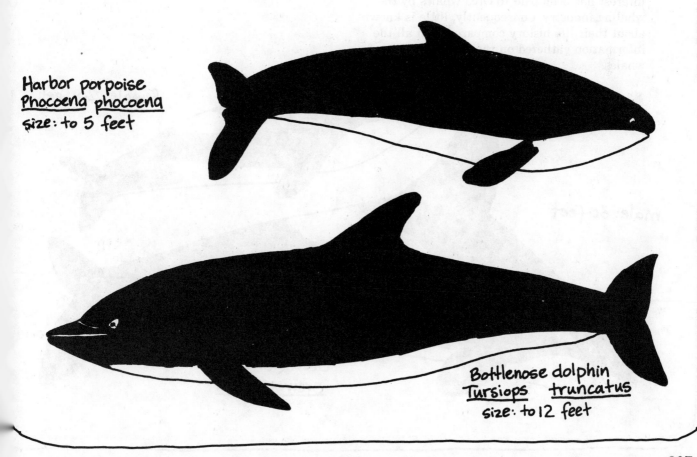

Harbor porpoise
Phocoena phocoena
size: to 5 feet

Bottlenose dolphin
Tursiops truncatus
size: to 12 feet

Orca—the Killer Whale

Orca whales spyhopping

Killer whales, *Orcinus orca*, are large dolphins that don't deserve the reputation the word "killer" suggests. True, Killer whales are voracious eaters and swallow fish, squid, seabirds, seals, and porpoises whole, but they need a lot of food to maintain their high activity level and large body. Contrary to misinformed sources, Killer whales do not kill for sport, only for food. The name Orca is preferred to Killer whale. In captivity, Orca whales are very gentle, trainable, affectionate, and accepting of people.

Orca whales usually have a striking black and white pattern, but totally black and totally white whales have been seen. Males have a six-foot-tall dorsal fin. Rarely are Orcas seen swimming alone. They generally stay in groups, with mature males swimming together, and females and immature males swimming together. Underwater, Orca whales have excellent vision. To see above the water, they "spyhop," that is, stand vertically (up to halfway) out of the water and look around.

On the Atlantic coast, Orca whales range from the Arctic to the Bahamas, but they are found throughout the world. Orcas are not thought to be in danger of extinction. Little interest has been paid to Orca whales by the whaling industry. Consequently, little is known about their life history compared with all the information gathered on the hunted species of whales.

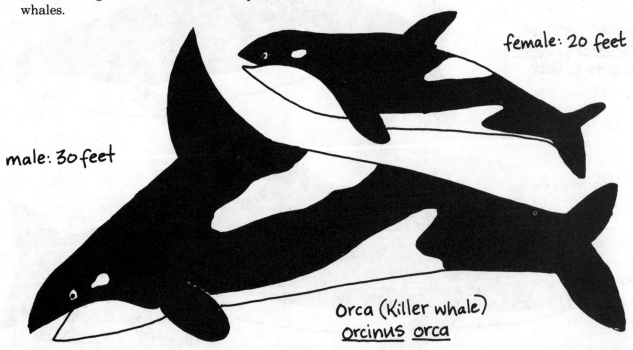

female: 20 feet

male: 30 feet

Orca (Killer whale)
orcinus orca

Seals

The term seal is often applied to any marine mammal in the order Pinnipedia (literally: those with winged feet), including the walrus, the sea lion, and the true seals. More narrowly, the term seal applies only to the true, or "earless" seals. Earless seals can hear, but they have no visible external ear. Sea lions and fur seals are considered eared seals. Sea lions are easy to teach and are often featured in juggling shows.

Only one of the seventeen species of seals in the world lives in warm water; the other sixteen live in cold and temperate water. There are no eared seals in the North Atlantic, but two species of true seals occur in the Gulf of Maine: the Harbor seal, *Phoca vitulina*, and the Gray seal, *Halichoerus grypus*. A few individuals stray as far south as the Carolinas.

Eared seals use their hind flippers to help them walk on land, but the true seals wriggle on their bellies, pulling themselves along with their front flippers.

Pinnipeds come up onto land or ice at breeding time. Harbor seals breed in loosely organized colonies. Mothers give birth to one pup, which is nursed on rich (50 percent fat) milk, which helps build up an insulating layer of blubber.

Although they are awkward on land, true seals are agile in the water, where they pursue fish, squid, and assorted invertebrates. They can remain submerged for fifteen minutes before coming up for a breath or air. While diving, their heart rate decreases drastically and their blood flow to extremities of the body is cut off to conserve the seal's energy and oxygen supply.

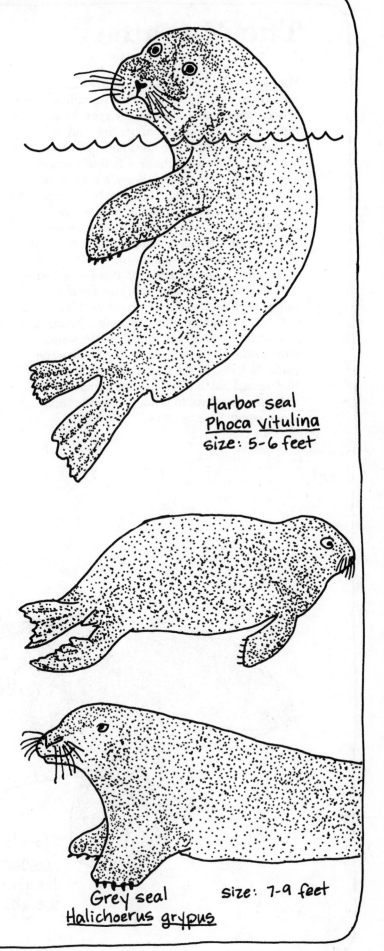

Harbor seal
Phoca vitulina
size: 5-6 feet

Grey seal
Halichoerus grypus

size: 7-9 feet

The Manatee

Manatees, *Trichechus manatus*, are sluggish, wrinkly marine mammals with thick lips and a bristly face. They have squinty eyes, broad front flippers and a flat tail. Sailors are said to have mistaken Manatees for mermaids, possibly because of poor eyesight or a fading memory of what women looked like. Female manatees, however, do nurse their baby by holding it up to their breast, much as human mothers do; perhaps from several miles off a desperate sailor could mistake a manatee for a mermaid, if he were imaginative enough.

Another name for the manatee is the Sea cow. This is quite an appropriate pseudonym, since the manatee is a voracious herbivore. In one day, a manatee may consume a hundred pounds of weeds, grass, and seaweed, which grow in the shallow warm waters where the manatee lives. In the past the state of Florida encouraged the introduction of manatees into clogged waterways to eradicate weeds that choke water flow.

Along the Atlantic Coast of the U.S., manatees once ranged from North Carolina to Florida. Now they are restricted almost exclusively to southern Florida, particularly the Everglades. By 1900, manatees were nearly extinct because of over-hunting. Their flesh, hide, and oil were fervently sought. In the wild, a manatee's natural predators are crocodiles and sharks. Having been designated an endangered species in Florida, manatees are now legally protected; there is a fine for harassing or molesting them. The biggest threats to manatees today (besides illegal shooting) are boat propellers and siltation of their feeding grounds. Channelization of southern Florida has drastically altered the run-off of water from land, and silt now accumulates where it never used to, closing the canals and filling in the manatee's habitats. Manatees lie placidly in shallow areas just under the surface, coming up to breathe every ten to fifteen minutes. All too frequently, they are hit by boats while submerged.

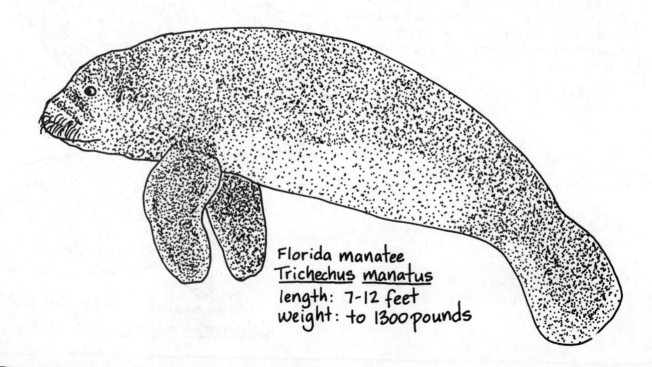

Florida manatee
Trichechus manatus
length: 7-12 feet
weight: to 1300 pounds

Marine Mammal True/False Quiz

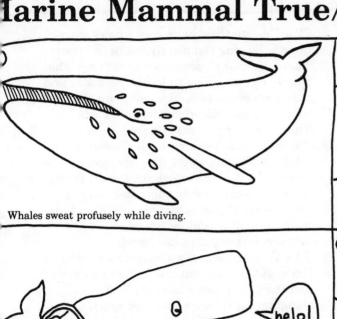

Whales sweat profusely while diving.

2. The Blue whale is the largest animal ever known on Earth.

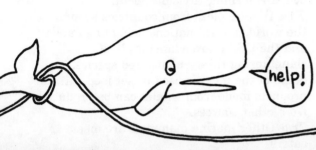

Whales can drown.

4. Whales can be told apart by the way they spout.

Whales never sleep.

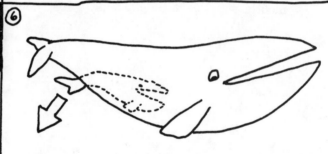

6. Baby whales are born tail first.

When whales breach, they jump completely out of the water.

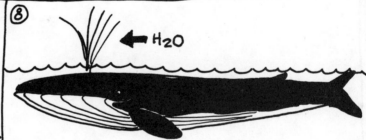

8. Whales spout by blowing water out their blowhole.

In Japan, dogs eat whale meat.

10. Manatees are the only vegetarian marine mammals.

1. *False* Whales have no sweat glands. When land mammals sweat, their body is cooled by the evaporation of the sweat. Evaporation doesn't occur underwater.
2. *True* Even the largest dinosaur is smaller than the Blue whale.
3. *True* Whales breathe air and will drown if they remain under water for too long. Sperm whales have become tangled in deep underwater telephone cables and drowned.
4. *True* Each species of whale has a distinctive spout. This helps in identifying whales from a distance. The Humpback spouts a low, rounded cloud; the Right whale's spout divides at the top.
5. *False* Although little is known of cetaceans' sleep patterns, they are known to doze at the surface, exhaling in a rhythmic pattern.
6. *True* Whale calves are born tail first. The mother immediately ushers her baby toward the surface for its first breath of air. The mother's nipples are located within a slit under her belly, and when the calf pushes against the slit, milk is sprayed into its mouth.
7. *True* Despite their huge size, whales can breach, leaping completely out of the water. There are many reasons for breaching. The whale may be trying to loosen parasites and barnacles on its skin. Breaching may be a form of communication with other whales. Then again, it just may be fun!
8. *False* The exhaled breath (the "spout") is a mixture of gases, water droplets, and a little mucus. As the whales exhale, the air from their lungs mixes with the water coming into the nasal passages. This mixture of air, water, and mucus often condenses in the cold air, forming a visible spout.
9. *True* Despite pleas from countries around the world, several nations, including Japan and the USSR, are adamant about slaughtering these endangered species. Some of the meat is used in pet food. All the products made from whales can be produced from other sources.
10. *True* Other marine mammals are meat eaters.

18 Marine Communities

A community is frequently defined as the living component of an ecosystem. A beach ecosystem, for example, consists of all the living organisms there (the beach community) plus the non-living components of the beach, such as the sand, sun, and surf. Some communities are named after their prodominant organisms: the "coral reef" community, the "eelgrass community" or the "mangrove" community. Others are named for their location: the "rocky shore" community or the "salt marsh" community.

Within a community, organisms interact with each other in various ways. Of primary importance is how they divide the available energy resources (food). Through photosynthesis, the plants in a community produce the food on which the animals depend. The animals in turn eat the plants or eat each other. Dead organisms are ultimately decomposed, releasing and recycling nutrients through the community.

Over half the population of the United States lives within 100 miles of the coast. The intertidal and shallow waters at the edge of the sea are very vulnerable to the effects of human activity. This is particularly true of estuarine communities, which are directly affected by the river runoff of populated upland areas. Although each community is composed of a unique assemblage of organisms, communities do not live isolated from one another. They exchange food and nutrients as well as pollutants and pesticides.

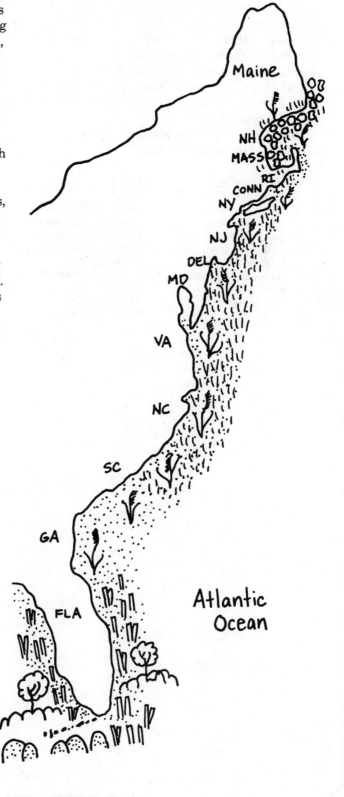

Atlantic Ocean

Rocky Shore Community

Sandy Beach Community

Salt Marsh Community

Eel Grass Community

Turtle Grass Community

Mangrove Community

Coral Reef Community

Fouling Community

For organisms living in the rocky intertidal zone, life is a constant challenge. As the tide comes in, or when storms occur, the intertidal organisms are hit by crushing waves. When the tide recedes, they are left high and dry. Basins among the rocks trap water from the outgoing tide, and these tidepools remain until covered by the next high tide. Organisms in tidepools must deal with extreme fluctuations in temperature, salinity, and oxygen availability. Because of their differing abilities to live out of the water, intertidal organisms appear in specific horizontal bands along the shore. Others that live near or below the low tide line can only stand to be exposed to the air for very short periods of time.

The following list corresponds to the organisms shown on page 215.

1. marine lichens — *Xanthoria, Verrucaria*
2. blue-green algae — *Calothrix*
3. Rough periwinkle — *Littorina saxatalis*
4. Acorn barnacle — *Balanus balanoides*
5. Green crab — *Carcinus maenus*
6. Atlantic plate limpet — *Acmaea testidinalis*
7. Atlantic dog whelk — *Thais lapillus*
8. Common periwinkle — *Littorina littorea*
9. Rockweed — *Fucus vesiculosus*
10. Blue mussels — *Mytilus edulis*
11. Coiled spiral worms — *Spirorbis borealis*
12. Knotted wrack — *Ascophyllum nodosum*
13. beach hopper — *Gammarus*
14. Northern red chiton — *Isochiton ruber*
15. Twelve-scale worm — *Lepidonotus squamatus*
16. Red-gilled nudibranch — *Coryphella verrucosa*
17. Green sea urchin — *Stronglycentrosus drobachiensis*
18. Irish moss — *Chondrus crispus*
19. Smooth periwinkle — *Littorina obtusata*
20. Dulse — *Palmaria palmata*
21. Clam worm — *Nereis virens*
22. Sea lettuce — *Ulva lactuca*
23. Rock eel — *Pholis gunnellus*
24. Frilled sea anemone — *Metridium senile*
25. kelps — *Alaria, Laminaria*
26. New England neptune — *Neptunea decemcostata*
27. sea squirts — *Molgula, Boltenia, Ciona*
28. Bread crumb sponge — *Halichondria panicea*
29. Rock crab — *Cancer irroratus*
30. sea lace — *Electra, Membranipora*
31. Lumpfish — *Cyclopterus lumpus*
32. Dead man's finger sponge — *Haliclona oculata*
33. Star tunicate — *Botryllus*
34. Eastern white slipper shell — *Crepidula fornicata*
35. Orange-footed sea cucumber — *Cucumaria frondrosa*
36. Sea star — *Asterias*
37. Daisy brittle star — *Ophiopholis aculeata*
38. Spiny sunstar — *Crossaster papposus*

Rocky Shore Community

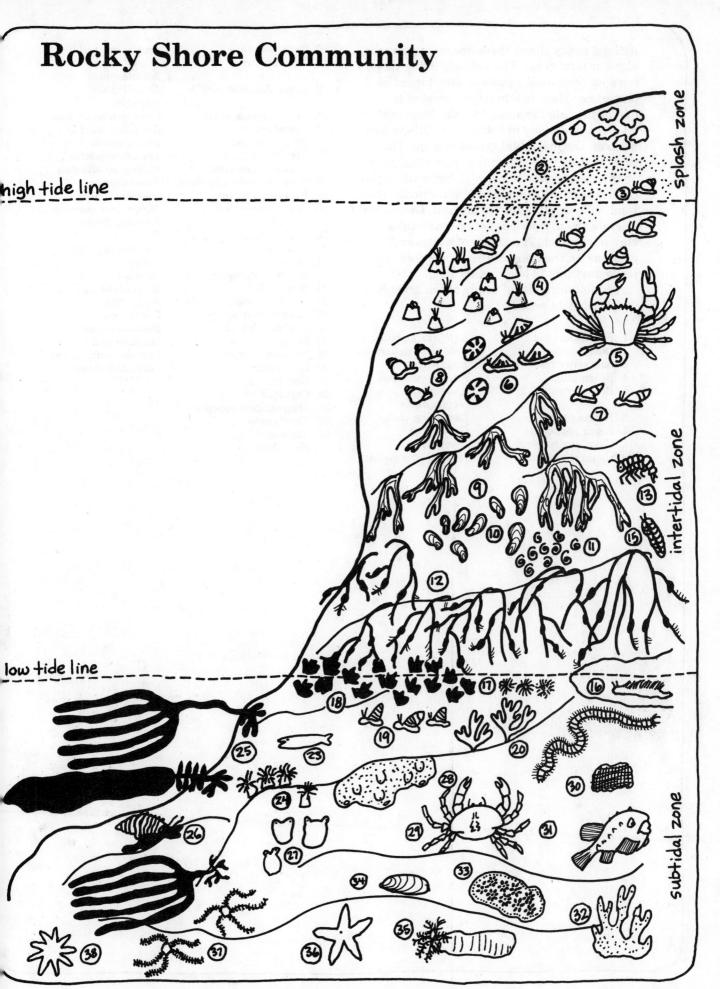

high tide line

low tide line

splash zone

intertidal zone

subtidal zone

Along a sandy shore, there are no large rocks, algae or tidal pools. The animals living there have no conspicuous place of attachment or protection. Most are therefore adapted to burrowing into the sand, like the Mole crab, the bivalves, and many of the worms. Others live between the individual grains of sand. These tiny creatures are called the *interstitial fauna* or the *meiofauna* (see page   ). Some intertidal creatures of the sandy beach move under the sand, following the tide in and out. Others can tolerate the dry periods between the tides and stay in one spot. The action of waves and currents on the ever-shifting sands does not leave anything in one spot for long.

The following list of sandy beach organisms corresponds to page 217:

1. Beach pea — *Lathyrus japonica*
2. Dusty miller — *Artemisia stelleriana*
3. Salt spray rose — *Rosa rugosa*
4. Beach plum — *Prunus maritima*
5. Beach grass — *Ammophila breviligulata*
6. Herring gull — *Larus argentatus*
7. beach hoppers — *Gammarus*
8. Clam worm — *Nereis virens*
9. Gould's sipunculid — *Phascolopsis gouldi*
10. Trumpet worm — *Pectinaria gouldi*
11. Acorn worm — *Saccoglossus kowalewski*
12. Giant Atlantic cockle — *Dinocardium robustum*
13. Milky ribbon worm — *Cerebratulus lacteus*
14. Lugworm — *Arenicola marina*
15. Soft shell clam — *Mya arenaria*
16. Florida coquina — *Donax variabilis*
17. Atlantic surf clam — *Spisula solidissima*
18. Atlantic jackknife clam — *Ensis directus*
19. Buttercup lucine — *Anodontia alba*
20. Northern quahog — *Mercenaria mercenaria*
21. moon snail — *Lunatia, Natica*
22. Fiddler crab — *Uca*
23. Mole crab — *Emerita talpoida*
24. sea star — *Asterias*
25. Summer flounder — *Paralichthys dentatus*
26. Sand dollar — *Echinarchnius parma*
27. terebellid worm — *Amphitrite*
28. Long-finned squid — *Loligo peali*
29. Knobbed whelk — *Busycon carica*
30. Sting ray — *Dasyatis centroura*
31. Horseshoe crab — *Limulus polyphemus*
32. Blue crab — *Calinectes sapidus*
33. Gastrotrich
34. Ostracod
35. Harpacticoic copepod
36. Tardigrade
37. Flatworm
38. Nematode

Sandy Beach Community

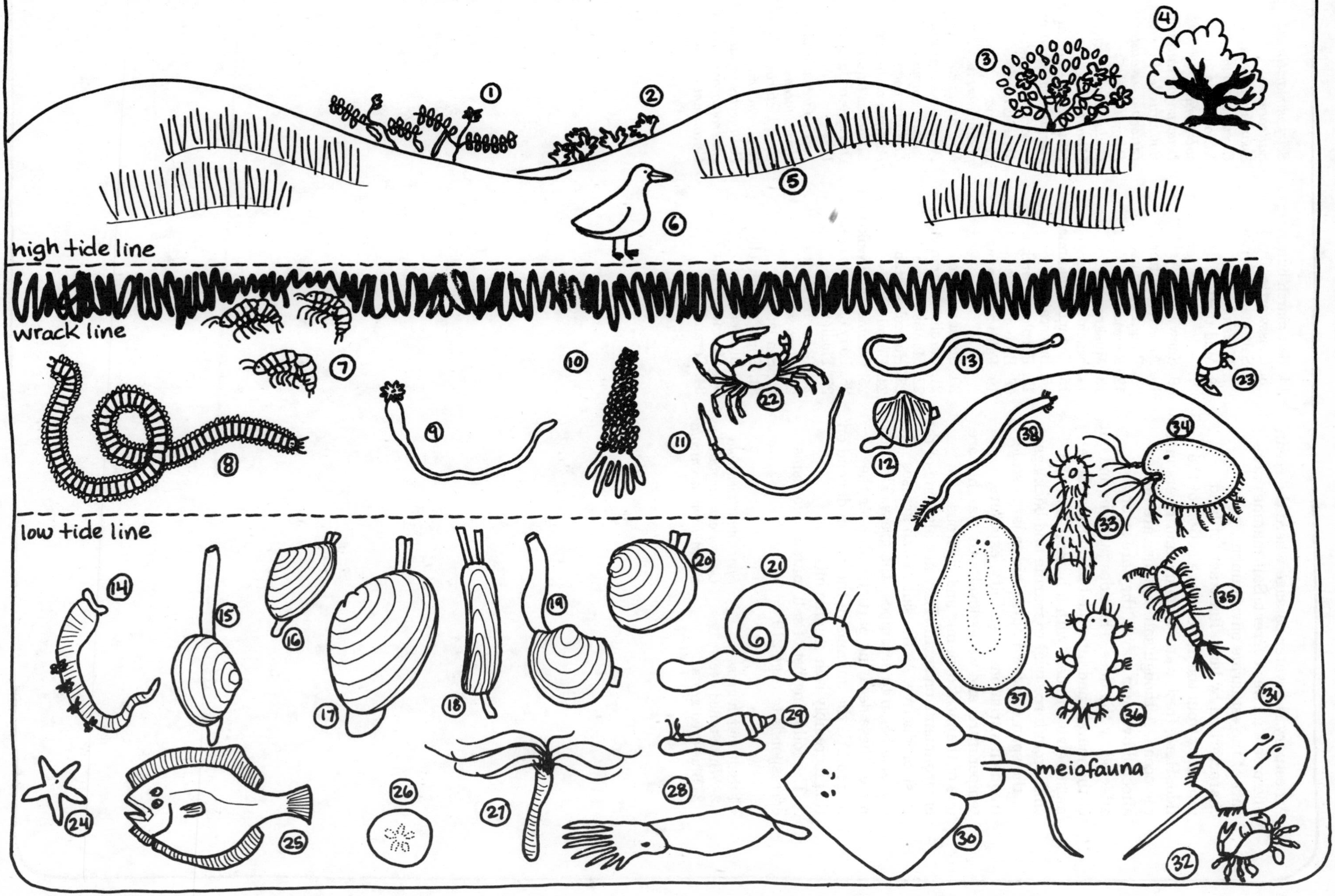

high tide line

wrack line

low tide line

meiofauna

217

An estuary is a water passage where the ocean tide meets a river current. Salt marshes are part of the estuarine environment, characterized by large flat tracts of land protected from the direct wave action of the tide. Here, fresh water streams and rivers meet the ocean. Although salt marshes are not affected by ocean waves, they are strongly affected by the tides. The tides pulse in and out of the marsh, and intertidal organisms there must be able to deal with a habitat that can shift from terrestrial to marine and back again in the span of a few hours. Due to large scale, frequent variations in salinity, exposure, temperature, and water level, salt marshes are one of the most dynamic and rigorous environments on earth. They are also one of the most productive. As the salt marsh grasses rot, bacteria begin to decompose the grass. The tide mixes these decaying particles all over the marsh. All the animals in the marsh depend on this incredibly rich organic mixture.

The following list of salt marsh organisms corresponds to page 219.

1.	Salt marsh grass	*Spartina alternifola*
2.	White-tailed deer	*Odocoileus virginianus*
3.	Blue-eyed grass	*Sisrinchium arenicola*
4.	Salt marsh bulrush	*Scirpus maritimus*
5.	Glasswort	*Salicornia*
6.	Salt meadow grass	*Spartina patens*
7.	Marsh hawk	*Circus cyaneus*
8.	Mud turtle	*Kinosternon subrubrum*
9.	Snowy egret	*Leucophyox thula*
10.	American bittern	*Botaurus lentiginosus*
11.	Rough periwinkle	*Littorina saxatalis*
12.	Raccoon	*Procyon loter*
13.	Muskrat	*Ondatia zibethicus*
14.	beach hoppers	*Gammarus*
15.	Common mummichog	*Fundulus heteroclitus*
16.	Bay scallop	*Aequipecten irradians*
17.	Purple sea urchin	*Arbacia punctulata*
18.	Blue mussels	*Mytilus edulis*
19.	False angel wing	*Petricola phaladiformis*
20.	Steamer clam	*Mya arenaria*
21.	Carolina marsh clam	*Polymesoda carolina*
22.	Mantis shrimp	*Squilla empusa*
23.	Horseshoe crab	*Limulus polyphemus*
24.	Hermit crab	*Pagurus pollicaris*
25.	Common periwinkle	*Littorina littorea*
26.	Lady or Calico crab	*Ovalipes cellatus*
27.	Fiddler crab	*Uca*
28.	Goose barnacles	*Lepas fascicularis*
29.	Clam worm	*Neries virens*
30.	Sea lettude	*Ulva lactuca*
31.	Beroe's comb jelly	*Beroe cucumis*
32.	Blue crab	*Callinectes sapidus*
33.	Mermaid's hair	*Enteromorpha erecta*
34.	Ceramium	*Ceramium rubrum*

Salt Marsh Community

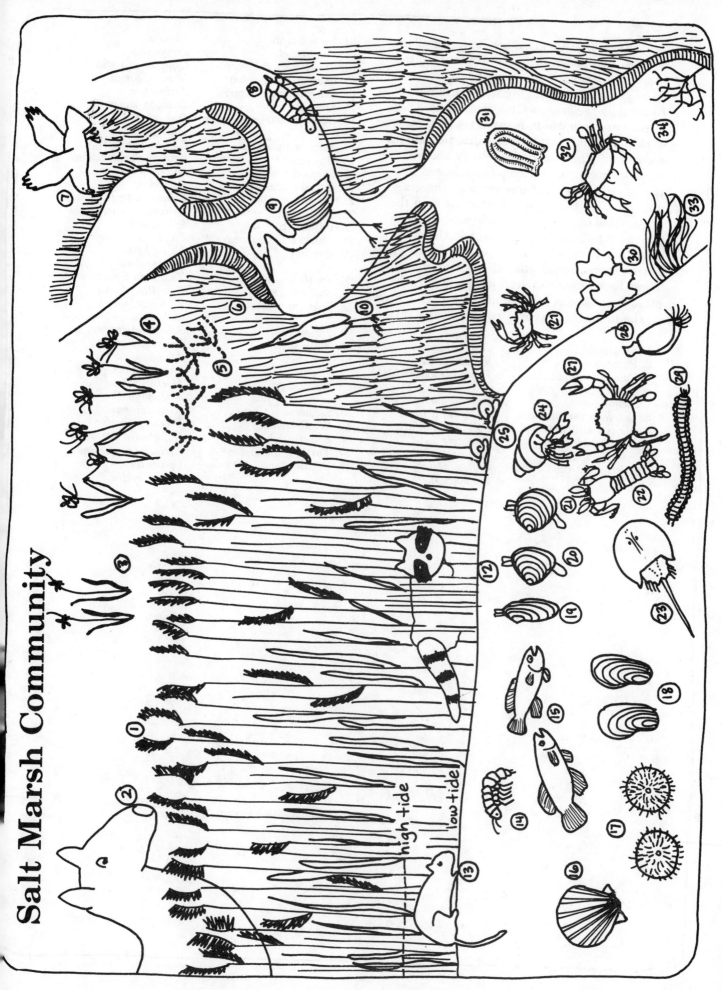

high tide

low tide

Eelgrass and Turtle grass are flowering plants adapted to relatively shallow water along the shore. The underground stems (rhizomes) of these grasses take hold in sandy bottoms. Eventually the grasses form dense beds. These grasses are valuable in many ways. The masses of rhizome hold down and stabilize the bottom sediments. When the grasses decay and are eaten by bacteria, nutrients are released into the water. Many creatures graze on the grass directly or on the organisms that grow on the grass. Dead blades that the tide washes up on the shore help stabilize and protect the shoreline. Beds of live Eelgrass and Turtle grass harbor large communities of marine organisms. These beds provide shelter, food, and sites of attachment for its inhabitants.

The following list corresponds to page 221.

1. Eelgrass — *Zostera marina*
2. Pipefish — *Sygnathus fuscus*
3. Common mummichog — *Fundulus heteroclitus*
4. Moon jelly (polypoid stage) — *Aurelia aurita*
5. Common periwinkle — *Littorina littorea*
6. Striped killifish — *Fundulus majalis*
7. Coiled tube worm — *Spirorbis*
8. Soft-shell clam — *Mya arenaria*
9. Parchment tube worm — *Chaetopterus variopedatus*
10. hermit crab — *Pagurus*
11. nudibranch — *Coryphella*
12. Sheepshead minnow — *Cyprinodon variegatus*
13. encrusting bryozoan — *Membranipora*
14. Burrowing anemone — *Actinothoe*
15. Sand shrimp — *Crangon septemspinosa*
16. Short-spined brittle star — *Ophioderma brevispina*
17. Slender flatworm — *Euplana gracilis*
18. Bay scallop — *Aequipecten irradians*
19. Plumed worm — *Diopatra cuprea*
20. Coiled spiral worm — *Spirorbis*
21. Decorator crab — *Stenocionops furcata*
22. sea urchin — *Lytechinus variegatus*
23. Pufferfish — *Sphaeroides*
24. Chicken liver sponge — *Chondrilla nucula*
25. Cowfish — *Lactophyrus quadricornis*
26. Turtle grass — *Thalassia testudinum*
27. Star tunicate — *Botryllus planus*
28. Colonial tunicates — *Ecteinascidia turbinata*
29. Fan worms — *Sabella melanostigma*
30. Tulip shell — *Fasicolaria tulipa*
31. Pen shell — *Atrina*
32. File fish — *Monocanthus*
33. Hydroids — *Macrorynchia philippina*
34. Seahorse — *Hippocampus*
35. White sponge — *Geodia gibberosa*
36. Sea cucumber — *Holothuria floridana*
37. Turtle grass foram — *Archaias angulatus*
38. Brittle star — *Ophiactis quinquerodia*

Eelgrass Community

Turtle Grass Community

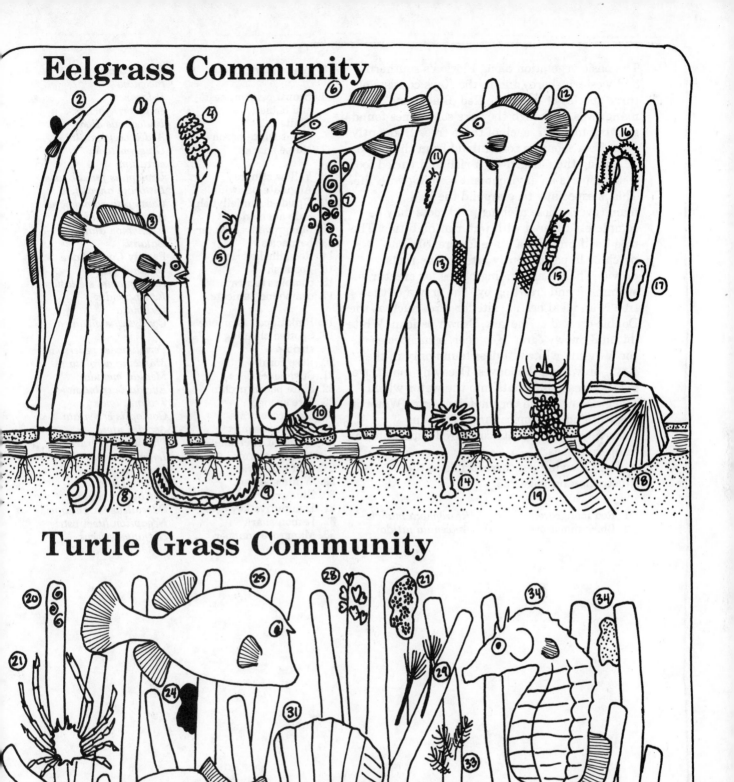

The basic vegetation along Florida's southern and western shores and in the Keys consists of three species of trees: the Red, Black, and White mangroves. Although they are sometimes found growing together, each of these trees frequently occupies a separate zone inland from the shore.

Among the mangroves, a rich and varied food web supports an environment teeming with life. This is especially true around the Red mangroves, whose arching prop roots may be completely covered with water at high tide. The tree itself provides a place for creatures to perch or to permanently attach. The roots provide shelter and hiding places for creatures swimming and crawling by. The crown of the tree is an ideal nesting site for birds such as the Double-crested cormorant, Brown pelican, White ibis and Snowy egret. Key deer, a subspecies of the Virginia white-tail deer found only in the Keys, browse on the leaves. Decaying mangrove leaves release nutrients into the water which otherwise would be very nutrient-poor. Water in tropical and subtropical areas contains relatively few nutrients compared with water of temperate seas.

The following list corresponds to the illustration on page 223.

1.	Red mangrove	*Rhizophora mangle*
2.	Black mangrove	*Avecennia nitida*
3.	White mangrove	*Laguncularia racemosa*
4.	Brown pelican, nest and babies	*Pelacanus erythrorrhynchus*
5.	White ibis	*Eudocimus albas*
6.	Great white heron	*Ardea occidentalis*
7.	Nurse shark	*Ginglymstoma cirratum*
8.	Yellow stingray	*Urolophus jamaicensis*
9.	Angulate periwinkle	*Littorina angulifera*
10.	Upside-down jellyfish	*Cassiopeia xamachana*
11.	Pale anemone	*Aiptasia pallida*
12.	Mangrove tree oyster	*Isognomon alatus*
13.	barnacles	*Balanus*
14.	Red-gilled worm	Family *Cirritulidae*
15.	Toadfish	*Opsanus tau*
16.	complex tunicate	*Amaroucium stellatum*
17.	compound tunicate	*Ecteinascidia turbinata*
18.	Pink-tipped anemone	*Condylactis gigantea*
19.	Spotted cleaning shrimp	*Periclimenes pedersoni*
20.	Rose coral	*Manicina aerolata*
21.	Silversides	*Menida menida*
22.	Slate pencil urchin	*Eucidaris tribuloides*
23.	Fire sponge	*Tedania ignis*
24.	Mangrove crab	*Goniopsis cruentata*
25.	Black sea squirt	*Ascidia nigra*
26.	Speckled anemone	*Phymanthus crucifer*
27.	Medusa worm	*Loimia medusa*
28.	Golf ball coral	*Favia fragum*
29.	White sponge	*Geodia gibberosa*
30.	Blue crab	*Callinectes sapidus*
31.	Diminutive key deer	*Odocoileus virginianus*
32.	Lemon shark	*Negaprion brevirostris*
33.	Mangrove snapper	*Lutjanus griseus*

Mangrove Community

A living coral reef is basically a sheet of living coral polyps which grow a few millimeters a year along with, and on top of their skeleton. Millions of organisms (up to 3,000 different species) congregate around a reef, for it provides plenty of hiding places and a supply of food. Few animals eat the coral itself, but they do eat each other. Many of the fish and invertebrates at the reef hide in the cracks and crevices by day and emerge only at night. At night, many nocturnal creatures venture out past the reef to graze on Turtle grass patches. Reef fishes often spend their juvenile stages in mangrove or Turtle grass areas, rather than on the reef itself.

Since the symbiotic zooxanthellae in the tissue of the coral polyps need sunlight for photosynthesis, coral reefs only grow in clear water and at depths where there is adequate light penetration.

The following list corresponds to the illustration on page 225.

1.	Purple jellyfish	*Pelagia noctiluca*
2.	Green turtle	*Chelonia mydas*
3.	Elkhorn coral	*Acropora palmata*
4.	Spanish hogfish	*Bodianus rufus*
5.	Great barracuda	*Sphyraena barracuda*
6.	Cocoa damselfish	*Pomacentrus variabilis*
7.	Comb jelly	*Menmiopsis mccradyi*
8.	Glassy sweeper	*Pempheris schomburgki*
9.	Neon goby	*Gobiosoma oceanops*
10.	Red-lip blenny	*Ophioblennius atlanticus*
11.	Mountainous star coral	*Montastrea annularis*
12.	File clam	*Lima scabra*
13.	Spotfin butterflyfish	*Chaetodon capistratus*
14.	Coconut macaroon urchin	*Tripneustes ventricosus*
15.	Flamingo tongue snail	*Cyphoma gibbosum*
16.	Common sea fan	*Gorgonia ventalina*
17.	Trumpetfish	*Aulostomas maculatus*
18.	Gray angelfish	*Pomacanthus arcuatus*
19.	Basket star	*Astrophyton muricatum*
20.	Green moray eel	*Gymnothorax funebris*
21.	Spotted drum	*Equetus punctatus*
22.	Long-spined sea urchin	*Diadema antillarum*
23.	Porkfish	*Anisotremus virginicus*
24.	Christmas tree worm	*Spirobranchus giganteus*
25.	Arrow crab	*Stenorhynchus seticornis*
26.	Nassau grouper	*Epinephelus striatus*
27.	Jolthead porgy	*Calamus bajanado*
28.	Grooved brain coral	*Diploria labrinthiformis*
29.	Spiny lobster	*Panuliris argus*
30.	Bristle worm	*Hermodice carunculata*
31.	Tube sponge	*Callyspongia vaginalis*
32.	Peacock founder	*Bothos lunatus*
33.	Nurse shark	*Ginglymostoma cirratum*
34.	Lettuce coral	*Agaricia lamarcki*
35.	Sea biscuit	*Clypeaster rosaceus*

Coral Reef Community

Fouling Community

After a certain period of time, nearly any man-made object put into the ocean will be covered with fouling organisms. Huge numbers of barnacle, bryozoan, and hydroid larvae will settle on and attach themselves to the object. This happens to docks, pilings, and buoys, and is especially a problem for ships.

Soon after an object is placed in the water, a film of bacteria, blue-green algae, and protozoans develops on it. Apparently the chemicals released by this film attract larvae of the larger fouling organisms. Once a few fouling organisms settle down, it is not long before they attract others so that a lush, diverse fouling community develops. Most members of the community live attached to the object itself. Other members of the fouling community crawl through the attached organisms or live on them.

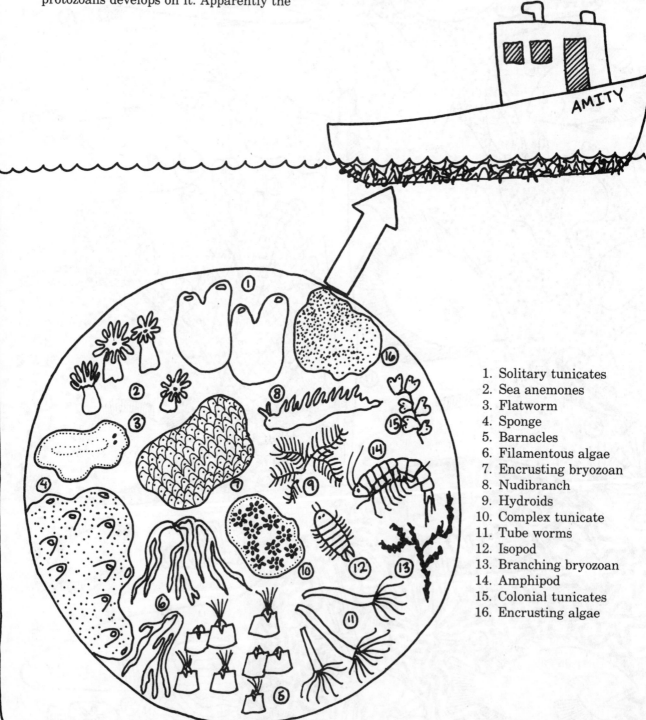

1. Solitary tunicates
2. Sea anemones
3. Flatworm
4. Sponge
5. Barnacles
6. Filamentous algae
7. Encrusting bryozoan
8. Nudibranch
9. Hydroids
10. Complex tunicate
11. Tube worms
12. Isopod
13. Branching bryozoan
14. Amphipod
15. Colonial tunicates
16. Encrusting algae

Marine Community True/False Quiz

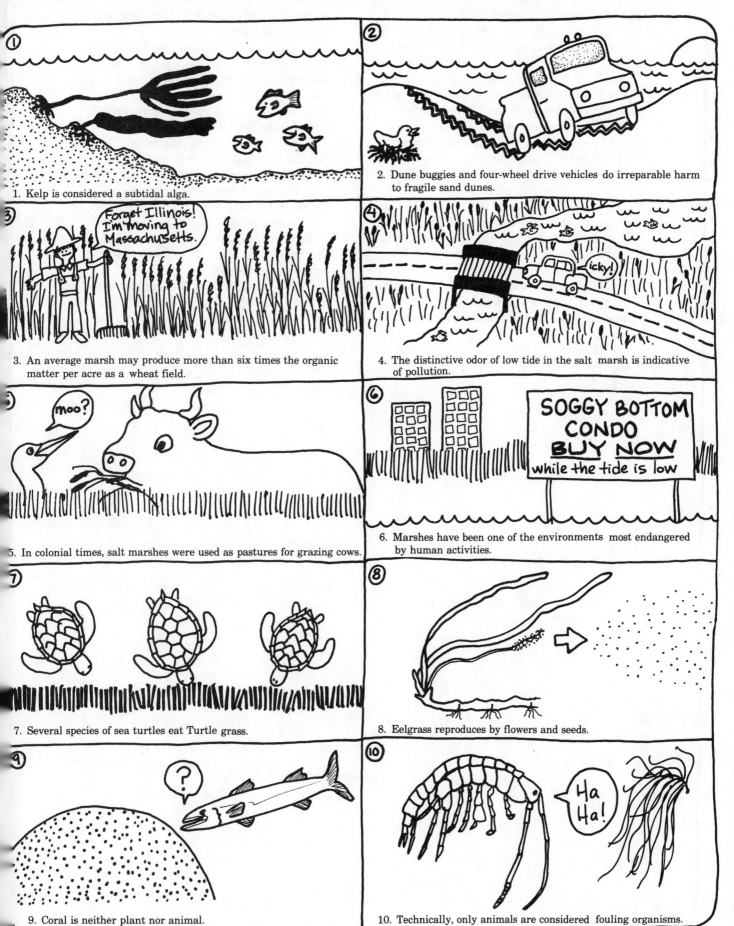

1. Kelp is considered a subtidal alga.

2. Dune buggies and four-wheel drive vehicles do irreparable harm to fragile sand dunes.

3. An average marsh may produce more than six times the organic matter per acre as a wheat field.

4. The distinctive odor of low tide in the salt marsh is indicative of pollution.

5. In colonial times, salt marshes were used as pastures for grazing cows.

6. Marshes have been one of the environments most endangered by human activities.

7. Several species of sea turtles eat Turtle grass.

8. Eelgrass reproduces by flowers and seeds.

9. Coral is neither plant nor animal.

10. Technically, only animals are considered fouling organisms.

1. *True* Although it is exposed by very low (spring) tides, kelp must remain submerged and is considered subtidal, not intertidal.

2. *True* A 1980 report from the President's Council on Environmental Quality said: "Off-road vehicles have damaged every kind of ecosystem found in the United States: sand dunes covered with American beachgrass on Cape Cod; pine and cypress woodlands in Florida; hardwood forests in Indiana; prairie grasslands in Montana; chaparral and sagebush hills in Arizona; alpine meadows in Colorado; conifer forests in Washington; arctic tundra in Alaska. In some cases the wounds will heal naturally; in others they will not, at least for millennia."

3. *True* Salt marshes are incredibly productive.

4. *False* The smell of low tide in the salt marsh is proof that the salt marsh system works. Rotting organisms (detritus) are the basis of the system.

5. *True* Salt marsh hay was also harvested.

6. *True* Unfortunately, marshland is often chosen as a site for land fill, dredging, oil refineries, and power plants.

7. *False* Green turtles are the only strictly herbivorous sea turtles.

8. *True* Eelgrass *is* a flowering plant. Its flowers are inconspicuous, and pollen is carried by the water. Eelgrass will flower once the water temperature exceeds 15°C. Between 10–15°C, it grows vegetatively (by rootstocks). Below 10°C, it is dormant.

9. *False* Coral is most definitely an animal. The individual animals (polyps) live together as a colony.

10. *False* All communities, fouling communities included, consist of animals and plants.

Geologic Time Scale

era	period	millions of years ago	life in the sea
PALEOZOIC	Cambrian	600	many types of algae and worms; trilobites abundant
	Ordovician	500	coral reefs begin; echinoderms well-established
	Silurian	425	(terrestrial life established) major groups, i.e. mollusks, continued to differentiate
	Devonian	405	continued differentiation; sharks abundant
	Mississippian	345	sharks dominant; first brittle star evidence
	Pennsylvanian	320	ancient coral groups waning; first reptiles appear
	Permian	280	drastic changes in climate; extinction of ancient stocks
MESOZOIC	Triassic	230	appearance of many modern invertebrate types
	Jurassic	180	dinosaurs abundant; birds appear; corals diversify
	Cretaceous	135	first record of dinoflagellates; planktonic forams abundant
CENOZOIC	Tertiary	63	most groups of modern plants and animals established
	Quaternary	1	Ice Ages; development of flora and fauna continues

Metric and Celsius Conversion

length
1 meter = 39.4 inches
1 centimeter = .39 inches

temperature

°C	°F	
0	32	freezing point of water
10	50	
20	68	
30	86	
40	104	
100	212	boiling point of water

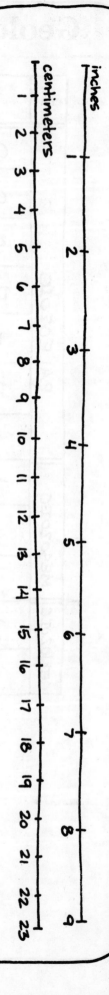

Glossary

abalone Edible snails (family Haliotidae) with a broad oval foot and a flat disk-like shell, which is a source of mother-of-pearl.

abdomen In arthropods and some annelids—the last of two or three major body divisions.

aboral Opposite or away from the mouth.

abyssal Pertaining to the great depths of the ocean, generally greater than 4,000 meters.

abyssopelagic Pertaining to the water above the bottom in the abyssal zone.

acorn worm Worm-like animals comprising the phylum Hemichordata.

Actinaria An order of coelenterates to which sea anemones belong.

adaptation Structural or functional changes of an organism in response to a new condition or environment; evolutionarily speaking, the organism becomes better suited to survive and reproduce due to these changes.

adductor muscle A muscle that closes or pulls together the shell of an organism—as the adductor muscles of bivalves or barnacles.

adipose fin In fishes, a fleshy fin usually lacking rays located behind (posterior to) the dorsal fin.

agar A gelatinous material obtained from certain red algae which is widely used for solidifying bacterial culture media in microbiology and as a stabilizer of emulsions.

air bladder In certain seaweeds such as the rockweeds, an airfilled sac which serves as a float; in fishes, the swim bladder.

Alcyonaria A subclass of coelenterates (synonymous with Octocorallia), which includes the soft corals and the horny corals; all alcyonarian polyps have eight pinnate tentacles.

algae (singular: alga) Unicellular or multicellular simple plants that have no vascular tissue and therefore no leaf, stem, or root systems.

algin A purified material obtained from brown seaweeds, extensively used for its thickening and emulsifying properties.

alternation of generations As in coelenterates, the alternation of a sexual generation (the medusa) with an asexual generation (the polyp).

ambergris Waxy material formed in the digestive tract of certain toothed whales, used in the making of perfume to fix odors.

ambulacral groove A groove on the oral surface of a seastar which contains the tube feet.

Amphineura A class of mollusks with a shell of eight calcareous plates; the chitons.

amphipod A crustacean of the order Amphipod including the beach hoppers and sand fleas.

amplitude The height of a wave crest or wave trough measured from the level of still water.

anadromous Migrating from seawater into freshwater to spawn, as salmon.

anal fin In fish, the unpaired fin behind the anus.

annelid Any member of the phylum Annelida, the segmented worms.

anterior Toward the forward or head end.

anus Opening at the end of the alimentary canal through which the waste products of digestion are excreted.

aperture An opening, such as the opening of a gastropod shell.

aphotic Without light, as in the aphotic zone of the ocean, that region below 600 meters where light is insufficient for plants to carry on photosynthesis.

aquaculture Cultivation of aquatic organisms.

Aristotle's Lantern A complex feeding structure of five teeth that surrounds the mouth of a sea urchin.

arrowworm Small transparent worm-like member of zooplankton, comprising the phylum chaetognatha.

Arthropoda A phylum of invertebrate animals characterized by jointed legs, a segmented body and an exoskeleton of chitin; includes the crustaceans and insects.

Aschelminth Somewhat heterogeneous phylum containing six classes of worms, including the rotifers, the gastrotrichs, the kinorhynchs, and the nematodes.

ascdian A member of the class Ascidiacea, a group of sessile tunicates (phylum Chordata) known as the sea squirts.

asconoid A type of sponge with a simple canal system in which the pores in the body wall lead directly to the internal cavity.

asexual Without sex.

asexual reproduction Reproduction that does not involve the union of sperm and egg.

atoll A circular coral reef surrounding a central lagoon.

autotrophic Pertaining to organisms which can synthesize organic products (food) from inorganic compounds, as do plants.

backbone The vertebral or spinal column.

bacteria Minute unicellular organisms, most of which are parasites or saprophytes; bacteria are the primary organisms responsible for decay and fermentation.

baleen The horny material that hangs in fringed plates from the upper jaw of baleen whales (also called whalebone), forming a sieve used to capture planktonic food.

bank An elevation of the sea floor surrounded by deeper waters, often excellent fishing areas.

barbel A slender fleshy sensory appendage near the mouth of various fishes such as catfish and Nurse sharks.

barnacle Sessile marine crustaceans comprising the order Cirrepedia.

barrier island A long, narrow, wave-built island parallel to the mainland.

barrier reef A coral reef parallel to the shore and separated from it by a lagoon. An example is the Great Barrier Reef of Australia.

basal disc In coelenterate polyps, the flattened adhesive disc at their base.

bathyal Pertaining to the benthic environment at depths of 200–4,000 meters.

bathypelagic Living in the water from depths of 1000–4000 meters.

benthic Living in or on the ocean bottom.

benthos The bottom of the sea or the organisms living there.

bilateral symmetry Having a body displaying two similar halves.

bioluminescence The production of light by living things.

binomial nomenclature A system of naming organisms in which each species is given a scientific name consisting of two words in Latin: the genus and the specific epithet. The genus name is capitalized, the specific epithet is not, but both are underlined or italicized, for example: *Callinectes sapidus*(the Blue crab).

biomass The weight of living organisms or organic material per area or volume.

Bivalvia The class comprised of mollusks with two-part shells, such as clams and oysters; class Pelecypoda.

black zone The upper limit of the intertidal zone where a band of blue-green algae grows.

blade The flat leaflike part of certain algae.

bloom Sporadic occurance of huge populations of algae.

blowhole Nostril on top of a whale's head through which it breathes.

blubber Fat of marine mammals.

blue-green algae Division Cyanophyta

body whorl The largest coil of a snail's shell.

bony fishes Members of the class Ostechthyes —those fish with an ossified (bony) skeleton.

boreal Northern.

boring sponge A sponge which bores into mollusk shells and rocks.

Brachyura An infraorder of decapods (Crustacea: Arthropoda) containing those decapods with their abdomen reduced and tucked tightly under the body; the true crabs.

brackish Salty, but less so than sea water.

breed To produce young; propagate.

brood To incubate.

brown algae Division Phaeophycophyta.

browse To feed on, nibble at, graze on.

Bryozoa A phylum of invertebrates containing the creatures known as moss animals and bryozoans.

buccal Of or pertaining to the mouth cavity.

bud An outgrowth, as on a sponge, capable of asexually developing into a new individual.

burrow A hole or passageway beneath the surface, or, to make such a hole.

byssus A filamentous structure secreted by certain mollusks with which they attach themselves to the substrate.

calcareous Containing calcium carbonate.

calcification Deposition of calcium carbonate within an organism.

calcium A soft metallic element found in nature, usually in combination with other elements.

calcium carbonate $CaCO_3$, a widely distributed compound occuring in nature as limestone and marble as well as being a component of invertebrate exoskeletons.

carapace The shield-like portion of exoskeleton that covers the head and thorax of various arthropods; also, the dorsal exoskeleton of a turtle.

carnivore A flesh-eating animal.

carrageenan A derivative of Irish moss, a red seaweed, used as an emulsifying agent.

caudal fin In fish, the tail fin.

cell The basic unit of which all living organisms are composed, usually consisting of a nucleus and a mass of cytoplasm bound by a membrane.

cellulose The primary constituent of the cell walls of plant cells.

centrifugal force An outward force on a body rotating about an axis.

Cephalopoda A class of mollusks containing the squids and octopi.

cerata Finger-like respiratory structures on the backs of nudibranchs.

Cetacea An order of marine mammals that includes the porpoises, dolphins, and whales.

Chaetognatha A phylum of small, planktonic invertebrates known as arrowworms.

chemoreceptor Structures or cells (taste cells, olfactory cells) that are stimulated by chemical substances.

chitin A nitrogenous carbohydrate (a protein) comprising the primary skeletal substance in arthropods.

chiton A mollusk in the class Amphineura.

Chlorophycophyta The division of plants comprising the green algae.

chlorophyll Green pigments in plants that absorb light energy needed in photosynthesis.

choanocyte In sponges, a cell with a flagella and collar of protoplasm.

Chordata The phylum comprised of animals that possess a notochord, gill slits, and a nerve cord at some stage of their life; includes the tunicates (subphylum Urochordata) and the vertebrates (subphylum Vertebrata).

cilia (singular: cilium) Minute hair-like projections.

ciliate A protozoan of the phylum Ciliophora.

ciliated Possessing cilia.

Cirripedia A subclass of crustaceans that includes the barnacles.

class A taxonomic group ranking below a phylum but above an order.

cleaner An animal that removes ectoparasites and dying tissue from the surface of another.

cloaca The end of the digestive tract that serves as a passageway for products of the reproductive, urinary, and digestive systems; not present in mammals.

Cnidaria A phylum of invertebrate animals that includes all the animals producing cnidocytes: hydroids, sea anemones, jellyfish and corals; also called Coelenterata.

cnidocyte A cell in coelenterates that contains the nematocysts, the stinging structures.

Coelenterata A phylum of invertebrate animals also known as Cnidaria.

coelenteron The digestive cavity of coelenterates.

cold-blooded: Having a body temperature that varies with the temperature of the surroundings.

colony A group of organisms of the same species living in close association with each other.

columella The spiral column in the center of a snail's shell.

comb jelly An animal belonging to the phylum Ctenophora; a ctenophore.

commensalism A symbiotic relationship where one partner benefits and the other is unaffected.

community A group of living organisms in a given area that interact with each other: the living component of an ecosystem.

competition The struggle among oranisms for food, space and other requirements for existence.

compressed Flattened from side to side.

consumer A heterotrophic organism; an animal.

continental drift The theory that continents are moving over the earth's surface.

continental shelf The sea floor adjacent to a continent extending from the shore to the depth of a marked increase in slope.

Copepoda An order of crustaceans, some of which are important members of the zooplankton.

copulation Sexual union of a female and male.

coral A coelenterate of the class Anthozoa, or sometimes the calcareous skeleton of an anthozoan.

coral reef A massive structure made of the calcareous skeleton of corals deposited over a long period of time.

coralline Pertaining to species of red algae that are heavily calcified.

core sample A vertical cylindrical sample of bottom sediments.

countershading The development of dark colors on parts of the body usually exposed to the sun and of light colors on parts of the body usually shaded.

cranium The part of a vertebrate's skull that encloses the brain.

crest The top of a wave.

Crustacea A class of arthropods that includes lobsters, crabs, and barnacles.

cryptic Pertaining to coloration that hides or conceals an organism.

Ctenophora The phylum that contains the comb jellies.

current A horizontal movement of water.

Cyanophyta A division of the plant kingdom containing the blue-green algae.

cyprid larva A free-swimming larval stage of barnacles.

Decapoda An order of crustaceans with five pairs of walking legs attached to the thorax. It includes lobsters, crabs, and shrimps.

decomposer Organisms, primarily bacteria, that break down dead organic matter into simpler substances.

demersal fishes Fish living near the bottom.

density Mass per unit volume of a substance; the density of seawater ($35\%_0$) at $0°C$ is 1.028 g/cm^3.

deposit feeding Feeding upon detritus that has settled to the ocean bottom.

depressed Flattened from top to bottom.

desiccation The process of being deprived of moisture.

detritivore An animal that feeds on detritus.

detritus Material resulting from the decomposition of dead organic remains.

diatoms Microscopic algae with a two-part siliceous cell wall; important members of the phytoplankton.

dinoflagellates Plant-like flagellated protozoans.

dioecious Unisexual; a dioecious organism is either male or female.

direct development Upon hatching, the young have the adult body form.

disphotic The dimly lit zone between the euphotic and aphotic zones without enough light to support photosynthesis.

dissolved oxygen Molecular oxygen present in water (*not* the 0 in H_2O).

diurnal tide A tide with one high water and one low water during a period of twenty-four hours, fifty minutes.

dorsal Pertaining to the back or upper surface of an animal's body.

dorsal fin In fish, the unpaired (median) fin (or fins) on the upper edge.

dorsoventrally From front to back.

dredge To remove sand, sediments, etc. from the bottom, using a scoop or shovel-like device.

ebb tide The movement of the tidal current away from shore; a decrease in the height of the tide.

Echinodermata Phylum of invertebrate animals including seastars, brittle stars, sea urchins, sand dollars, and sea cucumbers.

echolocation In cetaceans, the process by which they orient themselves by emitting high-frequency sounds.

elasmobranch The group of cartilaginous fish comprising the sharks and rays.

embryo A young organism in its early stages of development before it has emerged from the seed, egg or body of its mother.

embryology The study of an organism's early development.

emulsion A colloidal suspension of a liquid in another liquid.

encrusting To form a crust.

endoskeleton A skeleton that is produced within the body of an animal and remains embedded there.

epidermis In invertebrates, the outermost layer of cells.

epifauna Animals living on (but not in) the substrate.

epiflora Plants living on the substrate.

epipelagic Pertaining to organisms living in the upper 600 feet of the oceanic zone.

epiphyte A plant growing on another plant; more generally, any organism growing attached to a plant.

errant In polychaetes, those that are not confined to a burrow or tube.

estuary Brackish-water areas influenced by the tides where the mouth of a river meets the sea.

euphotic The upper layers of water (usually zero–600 feet) that receive enough light to support photosynthesis.

evert To turn outward or inside out.

evolution Process by which different types of organisms have developed from simpler forms through adaptation.

excrete To discharge.

exoskeleton An external skeleton, as the shells of mollusks and arthropods.

family A taxonomic group ranking below an order and above a genus.

feces Excrement.

fertilize Uniting of egg and sperm.

fetch Area of sea surface over which the wind blows with constant direction and speed, generating waves.

filter feed A type of supension feeding in which food particles are obtained by filtering them from a water current.

flagellate A protozoan of the phylum Mastigophora.

flagellum (plural: flagella) Slender whip-like structure on a cell.

flagellated Possessing flagella.

flatworm Any member of the phylum Platyhelminthes.

flora Plant life.

flowering plants Those plants that reproduce by means of flowers and seeds, as opposed to algae, which do not.

fluke 1. A parasitic flatworm, a trematode; 2. In whales, one of the horizontal tail lobes.

food chain The passage of energy (food) from producers (plants) up to herbivores and carnivores

food web Many interlocking and interdependent food chains.

foram Any protozoan of the order Foraminifera (phylum Sarcodina).

fouling Pertaining to organisms such as barnacles, bryozoans, and algae that live on the surface of man-made or introduced objects, usually in large numbers.

free-living Not parasitic or sessile.

fringing reef A reef attached directly to the shore of a continent or island.

fry Young fish.

gam A group of cetaceans swimming together.

gametes A mature sex cell, either sperm or egg.

Gastropoda A class of mollusks that includes the snails and nudibranchs.

genera Plural of genus.

generic name Name of a genus; the first word in a scientific name.

genus A taxonomic division of a family containing one or more closely related species.

Georges Bank A sandy shoal, 170 miles long and 90 miles wide, located in deep water off the coast of Massachusetts.

gill Respiratory structure in aquatic animals through which oxygen is extracted.

gill rakers In fish, projections on the inside of the gill arches that filter particles out of water passing over the gills.

gonad A reproductive organ; testis or ovary.

gorgonian A coral with a skeleton of horn-like material, gorgonin; the sea whips and sea fans (order Gorgonacea).

gravid With eggs; pregnant.

graze To feed on vegetation.

green algae Division Chlorophycophyta.

gribble Isopods that burrow into wood in salt water.

guano Phosphorus-rich excretion of seabirds.

gyre A circular or spiral form, usually pertaining to a current system in major ocean basins.

habitat Where an animal lives; its natural home.

halophyte A plant that can tolerate salty soils.

hard corals Corals that deposit a skeleton of calcium carbonate; those corals in the order Scleractinia.

Hemichordata A phylum of worm-like marine animals; the acorn worms.

hemoglobin Respiratory pigment containing iron found in the blood of vertebrates and in some invertebrates.

herbivore An animal that feeds entirely on plants.

hermaphrodite In animals, an individual with both male and female sex organs; one that can produce both sperm and egg.

hermatypic Pertaining to corals, those that possess zooxanthellae and are reef-builders.

heterocercal tail A type of fish tail in which the vertebral column extends into the upper lobe of the tail; usually the upper tail lobe is larger than the lower.

heterotroph An organism unable to manufacture its own food; an animal.

holdfast A structure of attachment, especially in algae.

holoplankton Organisms that never leave the plankton.

homocercel A type of fish tail in which both lobes are the same size and in which the vertebral column ends at or near the base of the tail.

horn A tough protein substance of keratin.

host An organism in which or on which another lives; in certain symbiotic relationships, the host is the larger of the two partners.

hydromedusae The medusoid (sexual) stage of hydrozoans.

Hydrozoa A class of Coelenterates which includes solitary polyps as well as colonial forms, such as the Portuguese man-of-war.

ichthyology The study of fishes.

immature Not fully developed.

indirect development Having a larval stage.

infauna Animals that live buried in the substrate.

ingest To take into the body, especially solid substances.

inorganic Not originating from an organism.

interstitial Pertaining to the organisms that live between sand grains or in other minute spaces.

intertidal The zone along the shore between high and low tide marks.

invertebrate Lacking a backbone; an animal without a backbone.

Isopoda The order of crustaceans that includes pill bugs and gribbles.

juvenile Immature; not fully developed.

kelp Large brown seaweeds in the order Laminariales.

keratin A fibrous protein present in structures such as claws, feathers and horn.

kingdom One of the five great divisions of living organisms: Monera, Protista, Fungi, Animalia, Plantae.

knot Unit of speed equivalent to one nautical mile per hour.

krill Common name applied to the shrimp-like crustaceans of the order Euphausiacea.

larva (plural: larvae) The immature form of an organism which does not resemble the adult.

lateral Pertaining to the side of an organism.

limestone Naturally occurring calcium carbonate.

littoral Pertaining to the seashore, especially the intertidal zone.

lophophore A ciliated O- or U- shaped feeding structure found in certain invertebrates, such as bryozoans.

lunate Crescent-shaped.

mackerel sharks Sharks in the family Lamnidae, including the Mako and Great white sharks.

macroplankton Zooplankton over one mm in size.

macroscopic Visible to the naked eye.

madreporite In echinoderms, a porous scab-like plate on the aboral surface through which water enters the water vascular system.

malacology The study of mollusks.

Mammalia A class of vertebrate animals characterized by mammary glands and hair.

mandibles In birds, the two halves of the beak. In other vertebrates, the upper portion of the jaw.

mangrove A general term applied to several species of tropical and sub-tropical salt-tolerant trees.

mantle In mollusks, the tissue layer which overlies the body and secretes the shell.

mariculture Aquaculture in the ocean.

marine pertaining to the ocean.

Mastigophora A phylum of protozoans with one or more flagella; the flagellates.

mate Joining together to breed.

mature Fully developed, adult.

mean Average, as in mean high tide or mean low tide.

medusa The jelly fish-shaped generation of coelenterates.

megalops larva Planktonic larval stage of crabs prior to the adult stage.

megaplankton Very large plankton, such as jellyfish, siphonophores and sunfish.

meiobenthos Small benthic organisms (100–500 microns) that live between sand grains.

meiofauna Same as meiobenthos.

meroplankton Temporary members of the plankton that spend part of their life as nekton or benthos.

mesoglea In coelenterates, the jelly-like layer between the epidermis and gastrodermis.

mesopelagic The intermediate depths between the euphotic and aphotic zones.

metabolism Energy changes which sustain life within an organism.

metamorphosis A change in form a animal undergoes as it develops from egg to adult.

metazoan Multi-celled animals; all animals excluding the phylum Porifera.

micron One thousandth (1/1000) of a millimeter, approximately equivalent to one twenty-five thousandth (1/25,000) of an inch.

microplankton Zooplankton between 60/1000 mm and one mm.

midrib The central vein of a leaf or the similar-looking structure of an algal blade.

milt Seminal fluid produced by male fishes.

mixed tide A tide with two highs and two lows per twenty-four hours and fifty minutes in which the two highs are unequal and the two lows are unequal.

Mollusca The invertebrate phylum containing gastropods, bivalves, and cephalopods.

molt To shed and regrow an exoskeleton or other outer body coverings.

Monera A taxonomic group, often considered a separate kingdom, that includes those organisms without true nuclei: the blue-green algae and bacteria.

monoecious Hermaphroditic; having both male and female sex organs in one individual.

moss animal Common name for members of the phylum Bryozoa.

mother-of-pearl Inner nacreous (pearly) layer of many mollusk shells.

motile Capable of moving or changing position.

mucus Slimy substance which serves to moisten and lubricate membranes, and to trap particles.

mutualism A symbiotic association where both partners benefit.

Mysticeti Baleen whales; a sub-order of the order Cetacea.

nacreous Composed of mother-of-pearl.

nannoplankton Minute planktonic organisms in the size range 5/1000–60/1000 mm.

nares Openings of the nasal cavity.

nauplius First planktonic larval stage of certain crustaceans.

neap tide Lowest range of the tide, occurring at the first and last quarter of the moon.

nekton Strong-swimming animals that can control their speed and direction.

nematocyst The stinging barb within cnidocytes of Coelenterates.

Nemertea The phylum containing the ribbon worms.

neritic Pertaining to water over the continental shelf, roughly the region from the low tide line to a depth of 600 feet.

neuston Minute organisms living in or at the surface film at the top of the water.

niche The place where an organism lives and the activities it carries out; its address and job.

nictitating membrane A thin membrane which functions as a third eyelid.

nitrogen-fixing Pertaining to organisms, especially bacteria and blue-green algae, that can process atmospheric nitrogen and convert it into compounds usable to other organisms.

nocturnal Active at night.

non-vascular Pertaining to plants without vascular tissue, such as algae.

notochord Primitive backbone.

nucleus Structure in plant and animal cells that contains genetic structures and coordinates the activities of the cell.

nudibranch A shell-less gastropod; a sea slug.

nutrients Substances which provide energy for growth and maintaining metabolism.

oceanic Zone of the pelagic environment where the water is deeper than 600 feet.

Octocorallia A subclass of of coelenterates comprising those species with eight pinnate tentacles; also called the subclass Alcyonaria.

Odontoceti Toothed whales; a suborder of Cetaceans.

olfactory Pertaining to smell.

omnivore An animal that eats both plants and animals.

operculum A lid or covering; in snails, the horny or calcareous structure that closes the shell aperture; in fish, the bony gill cover.

oral Pertaining to the mouth.

order The taxonomic group below a class and above a family.

organelle A structure within a cell that performs a specific function.

organic Derived from living organisms.

organism A living individual, plant, animal or otherwise.

osculum (plural: oscula) Excurrent pore of a sponge.

osphradium In mollusks, a sensory receptor that detects particles passing over the gills.

ossicles In echinoderms, tiny calcareous skeletal plates or fragments.

ossify To convert to bone or to develop a hard bone-like structure.

ostium (plural: ostia) An incurrent pore of a sponge.

ovary Female reproductive organ.

ovigerous Egg-bearing.

oviparous Pertaining to the type of reproduction in which eggs develop and hatch outside the female's body.

ovoviviparous Pertaining to the type of reproduction where young develop in eggs within the female's body and are then born alive.

palp In invertebrates, a small blunt appendage with various functions: sensory, feeding, etc.

Pangea A hypothetical continent early in the earth's history that contained all the continental crust of the earth.

parapodium (plural: parapodia) In polychaetes, a fleshy appendage that extends from the side of the body segment and functions in locomotion and respiration.

parasite An organism that lives on or in another organism, harming it in the process.

parasitism A symbiotic association in which one partner (the parasite) benefits and one partner (the host) is harmed.

patch reef An isolated coral growth, usually landward of a barrier or fringing reef.

peanut worm Common name for members of the phylum Sipunculida.

peat Partially decomposed vegetation.

pectoral fins In fish, the paired lateral fins near the gill cover.

pedicellaria In echinoderms, small pincer-like structures on the body surface.

pelagic The division of the ocean that includes the whole mass of water; it is divided into the neritic zone (water depth 0–600 feet) and the oceanic zone (water deeper than 600 feet).

Pelecypoda Another name for the class Bivalvia.

pelvic fins In fish, the paired fins on the ventral side.

pentamerous Divided into five parts.

pereiopod A walking leg attached to the thorax of a crustacean.

period of a wave Time between the passage of two successive wave crests past a certain point.

periostracum The thin black, flaky covering of a mollusk shell.

Phaeophycophyta The division of plants comprising the brown algae.

pharynx In invertebrates, the part of the alimentary canal connecting the mouth with the esophagus.

photic Pertaining to the layer of water that receives enough light to support photosynthesis.

photosynthesis The process in plants in which organic products (simple sugars: food) are formed from carbon dioxide and water in the presence of sunlight and chlorophyll.

phycology The study of algae.

phylogeny The evolutionary history of an organism.

phytoplankton Plant plankton.

pigment A substance that gives color.

pinnate Feather-like.

Pinnipedia Suborder of the order Carnivora that includes these aquatic mammals: seals, sea lions and walruses.

piscine Pertaining to fish.

placenta In mammals, the structure through which the fetus obtains food and discharges waste.

plankter An individual member of the plankton.

planktivore An animal that feeds on plankton.

plankton Aquatic organisms that float at the mercy of the currents or have limited swimming abilities.

plastron The bottom (ventral) part of a turtle's shell.

Platyhelminthes The phylum of invertebrates comprising the flatworms.

pleopods Abdominal appendages of certain crustaceans.

pneumatophore 1. In siphonophores, a modified zooid that forms a gas-filled float; 2. In mangroves, a breathing root.

podium (plural: podia) The tube foot of an echinoderm.

Polychaeta A class of annelid worms with well-developed parapodia.

polymorphism Having more than one form.

polyp In coelenterates, a sessile individual with a cylindrical body attached at one end and with a mouth at the other.

Polyzoa Another name for the phylum Bryozoa.

population All individuals of one species living in a particular area.

Porifera The invertebrate phylum comprising the sponges.

posterior At the back (tail, rear) end of the body.

ppt: Parts per thousand.

prawn Edible shrimp, especially those of the genera *Penaeus* and *Palaemonetes*.

predaceous Predatory.

predation The act of killing other animals for food.

predator An animal that kills other animals for food.

prey An animal killed for food.

primitive Possessing traits characteristic of ancestral species.

proboscis Elongated, tubular, extensible structure.

producer An organism that can produce organic substances from inorganic ones; plants.

prolific Producing many children.

propagate To breed.

prop root In mangroves, aerial root holding the tree erect.

protandry Type of hermaphroditism in which the individual first produces sperm, then eggs.

Protista A major taxonomic group, often considered a kingdom, that comprises most of the algae.

protoplasm The primary substance of a living cell.

Protozoa An invertebrate phylum comprising all unicellular animals.

pseudopodium (plural: pseudopodia) An extension or projection of the protoplasm of a cell, as in some protozoans.

pteropod An opistobranch gastropod with wing-like flaps; a sea butterfly.

p.s.i. Pounds per square inch; a measurement of pressure.

pulmonate Having lungs or lung-like structures.

Pycnogonidia A class of arthropods comprising the sea spiders.

Pyrrophyta A division of microscopic algae possessing flagella for locomotion; the dinoflagellates.

radial symmetry Having similar parts radiating from a central point.

Radiolarian An order of marine protozoans.

radula In mollusks, a tongue-like toothed structure that is used in rasping and chewing.

raptorial Adapted for seizing prey.

red algae Division Rhodophytophyta.

red tide A bloom of dinoflagellates, especially of the genera *Gonyaulax* and *Gymnodinium*, which, if present in large enough quantities, can color the water red.

reef An offshore ridge of materials such as rocks or coral that lie close to the surface of the water.

regeneration In invertebrates, the regrowth of a missing part or the restoration of a new individual from part of the original.

requiem sharks Sharks in the family Carcharhinidae.

respiration 1. within a cell, the liberation and utilization of energy; 2. the exchange of gases between an organism and its environment; 3. the act of breathing.

rhizome In plants, a horizontal stem on or under the ground that produces stems and roots; in animals, a horizontal outgrowth that gives rise to new individuals.

Rhodophycophyta A division of plants comprising the red algae.

ribbon worm Members of the invertebrate phylum Nemertea.

rockweeds Intertidal and subtidal brown seaweeds, especially of the genera *Fucus* and *Ascophyllum*.

roe The eggs of a fish, especially when still in the ovary.

rookery A breeding ground of gregarious birds or mammals.

roost A place where birds rest at night, often in large numbers.

rorqual Baleen whales of the genus *Balaenoptera*.

rostrum In crustaceans, a forward projection of the carapace beyond the head.

salinity The saltiness of the water measured in parts per thousand (abbreviated ppt or $^o/oo$).

salp A pelagic tunicate.

salt NaCl, sodium chloride; more generally, salts are any substance that yields ions other than hydrogen and hydroxyl.

salt marsh An area of soft wet land periodically flooded by salt water.

Sarcodina The phylum of protozoans including amebas and forams.

Sargassum A genus of brown algae, several species of which form most of the vegetation of the Sargasso Sea.

scavenger An animal that feeds on dead animals or dead organic matter.

school Many similar aquatic organisms swimming together.

scientific name See "binomial nomenclature."

scrod Young cod.

Scyphozoa A class of coelenterates comprising the jellyfish.

sea squirt A sessile tunicate.

seawater Water with an average salinity of 35 ppt.

seaweed Macroscopic marine algae, most of which grow attached to the substrate.

sedentary Permanently (or usually) attached to the substrate.

sediment The material that settles through the water column to the bottom.

seed In flowering plants, an embryo covered by a seed coat.

seismic Pertaining to or produced by earthquakes or vibrations of the earth.

self-fertilization The union of the egg and sperm of one individual.

semidiurnal tides Two similar high waters and two similar low waters during a period of 24 hours and 50 minutes.

sensory cells Cells specialized to receive stimulation from light, odors, tastes, sounds, etc.

sessile Sedentary, attached.

seston All organisms or matter that swims or floats in the water.

seta (plural: setae) Stiff bristle-like structures.

sexual reproduction Reproduction involving sperm and egg.

shell Hard exoskeleton of certain animals, especially mollusks and marine arthropods.

shellfish Aquatic vertebrates with a shell, particularly the edible varieties.

shoal Shallow.

silica SiO_2, silicon dioxide; occurs in skeletons of diatoms and sponges; it is the principal component of sand.

siliceous Containing silica.

silt Particles intermediate in size between sand and clay.

Siphonophora An order of coelenterates including floating hydrozoans, such as the Portuguese man-of-war.

Sipunculida A phylum of marine invertebrates known as the peanut worms.

skeleton The protective covering or inner framework of an organism.

slack water A tidal current with no velocity, usually occurring when the tide changes its flow.

soft coral Loosely applied to all corals in the subclass Octocorallia; those corals without a skeleton of calcium carbonate.

solitary Living alone, not colonial.

solubility The ability to be dissolved.

species A basic taxonomic group consisting of individuals of common ancestry who strongly resemble each other physiologically and who interbreed, producing fertile offspring; a division of a genus.

specific epithet The second half of a scientific name; it is not capitalized.

spicules Siliceous or calcareous skeletal elements of sponges and soft corals.

spiracles A pair of holes in the head of some sharks and rays through which water is drawn in and passed over the gills.

spongin Fibrous, horny, flexible material making up the skeleton of some sponges.

spore An asexual reproductive structure that can give rise to a new individual.

spray zone Zone above the high tide line that is regularly wet by the salt spray of the surf.

spring tide Tide of maximum range occurring at the new and full moon.

statocyst In some invertebrates, an organ of equilibrium.

stenohaline Having a low tolerance for changing salinity.

stenothermal Having a low tolerance for changing temperatures.

sterile Not fertile.

stinging cell In coelenterates, cells (cnidocytes) that contain stinging structures (nematocysts).

stolon A horizontal shoot which gives rise to new individuals.

stomata Small openings in a leaf or stem through which gases pass.

sublittoral Pertaining to the zone below the low tide line.

subtropical Nearly tropical in location or climate.

subspecies A subdivision of a species consisting of individuals different from the rest of the species but who can still interbreed with other members of the species.

substrate Solid material upon which an organism lives or to which it is attached; in the ocean this can be rocks, man-made structures, the sediment, or living organisms.

subtidal Pertaining to the zone below the low tide line.

surf Collective term for breakers.

suspension feeding Feeding upon particles, either plankton or detritus, suspended in the water.

swell Waves that have traveled away from the area in which they were generated.

swim bladder A hydrostatic gas sac in fish.

swimmeret A pleopod.

symbiont One of two partners in a symbiotic association.

symbiosis An association in which two dissimilar organisms live closely together.

symbiotic Pertaining to symbiosis.

systhesis The production of a compound from simpler constituents.

tactile Pertaining to the sense of touch.

talon A sharp claw.

taxonomy The science of classifying organisms into groups.

teleost Fish with a bony skeleton, the class Osteichthyes.

temperate zone That part of the earth's surface between the tropics and the poles.

test A covering on various animals, including the skeleton of some echinoderms and the tunic of tunicates.

thallus A simple plant body, as of an alga; a plant without roots, stems, or leaves.

thoracic Pertaining to the thorax.

thorax In invertebrates, the region of the body between the head and abdomen.

tidal day The time interval between two successive transits of the moon over a meridian; 24 hours and 50 minutes.

tidal range The difference in height between consecutive high and low tides.

tidal wave Tsunami, or huge sea wave caused by an ocean disturbance.

tide The periodic rise and fall of the sea level along coasts resulting from the tide-generating forces of the moon and sun on the earth.

tidepool Depression in a rock (or created by rocks) within the intertidal zone that traps water as the tide recedes.

tissue Cells of similar structure that are grouped together and perform a specific function.

toxin A poisonous substance produced by a living organism.

trawl A sturdy bag or net that is hauled along the bottom to catch bottom-dwelling animals.

trilobite An extinct group of primitive marine arthropods.

trocophore Ciliated planktonic larval stage of many invertebrate groups, including the mollusks and annelids.

tropics The region between the Tropic of Cancer and the Tropic of Capricorn.

trophic Pertaining to nutrition, food or growth.

tsunami A long-period wave produced by a submarine disturbance; a "tidal" wave.

tube foot The thin tubular external extensions of an echinoderm's water vascular system; a podium.

tunic The covering (test) of a tunicate.

tunicate Any member of the subphylum Urochordata; commonly called a sea squirt.

turbid Cloudy due to suspended matter in the water.

unicellular Composed of one cell.

univalve A mollusk with a one-piece shell; a gastropod.

Urochordata The subphylum of chordates containing the tunicates.

uropods In crustaceans, flattened appendages forming the tail fan, as in a lobster.

upwelling The process by which water rises from a lower to a higher depth.

valve In invertebrates, a distinct piece of a shell.

vascular Pertaining to or possessing vessels through which fluid travels.

vegetative reproduction In plants, asexual development.

veliger Planktonic larva of some mollusks.

velum Thin, shelf-like structure which projects inward from the margin of the bell in hydromedusae.

ventral Pertaining to the lower or underneath or abdominal surface of an animal.

vertebrate Any member of the subphylum Vertebrata (phylum Chordata) comprising the fish, birds, reptiles, amphibians and mammals.

viable Capable of growing or living.

viscera A collective term referring to the internal organs of an animal, especially those of the abdominal area.

viviparous Giving birth to live young attached to the mother by a placenta.

warm-blooded Having a constant and internally controlled body temperature.

waste Material eliminated from the body.

water-vascular system In echinoderms, the system of fluid-filled canals, often ending in tube feet, used in feeding and locomotion.

wave A disturbance that moves over the surface of or through the water.

wave height Vertical distance between a crest and the next trough.

whale bone Baleen.

Zoantharia The subclass of coelenterates containing the sea anemones and hard corals.

zoea larva Stage in the development of some crustaceans.

zooid An individual member of a colony.

zooplankton Animal plankton.

zooxanthellae Dinoflagellates that live symbiotically in an encysted state within certain invertebrates, especially coelenterates.

Bibliography

Abbott, R.T. *American Seashells*. New York: Van Nostrand Reinhold Co., 1974.

——————————— *Seashells of North America*. New York: Golden Press, 1968.

Barnes, R.D. *Invertebrate Zoology*. Philadelphia: Saunders College/Holt, Rinehart and Winston, 1980.

Berrill, M., and Berrill, D. *A Sierra Club Naturalist's Guide: The North Atlantic Coast*. San Francisco: Sierra Club Books, 1981.

Bigelow, H.B., and Schroeder, W.C. *Fishes of the Gulf of Maine*. Washington, D.C.: US Government Printing Office, 1953.

Carson, R. *The Edge of the Sea*. Boston: Houghton, Mifflin Co., 1955.

Dawson, E.Y. *Marine Botany*. New York: Holt, Rinehart and Winston, Inc., 1966.

Gosner, K.L. *A Field Guide to the Atlantic Seashore*. Boston: Houghton Mifflin Co., 1978.

Greenberg, I., and Greenberg, J. *Guide to Fishes of Florida, the Bahamas and the Caribbean*. Miami: Seahawk Press, 1977.

Gross, M.G. *Oceanography: A View of the Earth*. Englewood Cliffs, New Jersey: Prentice-Hall, Inc., 1972.

Katona, S., Richardson, D., and Hazard, R. *A Field Guide to the Whales and Seals of the Gulf of Maine*. Bar Harbor, Maine: College of the Atlantic, 1977.

Lagler, K.F., Bardach, J.E., and Miller, R.R. *Icthyology*. Ann Arbor: University of Michigan Press, 1962.

Levinton, J.S. *Marine Ecology*. Englewood Cliffs, New Jersey: Prentice-Hall, Inc., 1982.

McConnaughey, B.H. *Introduction to Marine Biology*. St. Louis: C.V. Mosby, Co., 1974.

Meglitsch, P. *Invertebrate Zoology*. New York: Oxford University Press, 1967.

Pough, R.H. *Audubon Water Bird Guide*. Garden City, New York: Doubleday and Company, Inc., 1951.

Robbins, S.F., and Yentsch, C. *The Sea Is All About Us*. Salem, Massachusetts: Peabody Museum of Salem, 1973.

Scientific American, Inc. *The Ocean*. San Francisco: W.H. Freeman and Company, 1969.

Schmitt, W.L. *Crustaceans*. Ann Arbor: University of Michigan Press, 1965.

Slipjer, E.J. *Whales and Dolphins*. Ann Arbor: University of Michigan Press, 1976.

Smith, D.L. *A Guide to Marine Coastal Plankton and Marine Invertebrate Larvae*. Dubuque, Iowa: Kendall/Hunt Publishing Co., 1977.

Smith, F.G.W. *Atlantic Reef Corals*. Coral Gables, Florida: University of Miami Press, 1971.

Smith, R.I. *Keys to Marine Invertebrates of the Woods Hole Region*. Woods Hole: Marine Biological Laboratory, 1964.

Thurman, H.V. *Introductory Oceanography*. Columbus: Ohio: Charles E. Merrill Publishing Co., 1975.

Ursin, M.J. *Life In and Around the Salt Marshes*. New York: Thomas Y. Crowell Co., 1977.

Voss, G.L. *Seashore Life of Florida and the Caribbean*. Miami: E.A. Seeman Publishing, Inc., 1976.

Index